N° 81 | 2023

# POLITIQUE EUROPÉENNE



# EU Digital Policies and Politics

Sous la direction de Céleste Bonnamy,
Clément Perarnaud

L'Harmattan

*Revue publiée avec le soutien du laboratoire Pacte Grenoble, de l'European School of Political and Social Sciences (ESPOL), de l'Institut des sciences humaines et sociales du CNRS, et de l'Institut d'études européennes (IEE) de l'Université Saint-Louis – Bruxelles*

Pacte
Laboratoire de sciences sociales
http://www.pacte-grenoble.fr/

ESPOL
EUROPEAN SCHOOL OF POLITICAL AND SOCIAL SCIENCES
www.espol-lille.eu

iee
Institut d'études européennes
UNIVERSITÉ **SAINT-LOUIS**-BRUXELLES

Politique européenne
Centre d'études européennes de Sciences Po
28, rue des Saints-Pères
F - 75 007 Paris
Tél. (+ 33 1) 45 49 83 52
Fax (+ 33 1) 45 49 83 60
http://politique-europeenne.eu/


www.editions-harmattan.fr
ISBN : 978-2-336-42866-6
EAN : 9782336428666

## *DOSSIER*



## *LECTURES CRITIQUES*

# Dossier

**POLITIQUE EUROPÉENNE**
**N° 81 | 2023**
Céleste Bonnamy, Clément Perarnaud
p. 8-27

# Introduction

# EU Digital Policies and Politics

## Unpacking the European Approach to Regulate the "Digital"

**Céleste Bonnamy**
Luiss Guido Carli (Rome, Italie)/CEVIPOL (Université libre de Bruxelles, Belgique)

**Clément Perarnaud**
Vrije Universiteit Brussel (VUB, Belgique)

## Introduction

In May 2022, the European Union's (EU) attempts to regulate the global influence of tech giants were vividly illustrated when Thierry Breton, European Commissioner for the Internal Market, engaged in a high-profile meeting with Elon Musk. The American entrepreneur, who had recently acquired the social media platform Twitter, now known as "X", met with Commissioner Breton to discuss the compatibility of the forthcoming Digital Services Act (DSA) with his own digital vision.[1]

Fast forward to October 2023, the DSA had transformed from draft law to enforceable legislation. The European Commission launched an investigation against X for the dissemination of illegal content, provoking a response from Elon Musk, who threatened to deactivate the social media platform in Europe. This episode exemplifies what we term the EU's "digital assertiveness" (Bendiek and Römer, 2019) on the global stage, emblematic of the evolving landscape of EU digital politics that this special issue seeks to explore.

Despite its initial lack of formal competence in digital policy (Mărcuț, 2017), the EU has been progressively developing a framework of legislative texts designed to regulate the digital realm.[2] These initiatives represent a crucial

1 The European Commission published a short video featuring Elon Musk and Thierry Breton discussing the DSA. Interestingly, the Commission broadcasted the video on its own official Twitter account. See: <https://twitter.com/UEFrance/status/1523937825921581056>.

2 To name a few: the 2016 General Data Protection Regulation (GDPR), the 2019 Directive on Copyright in the Digital Single Market (DSM), the 2022 Digital Services Act (DSA) and Digital Markets Act (DMA), the 2023 Digital

phase in the evolution of EU digital policy-making (Newman, 2020) and politics, reflective of the European Commission's commitment to shaping the digital realm since the early 2010s. This dynamic accelerated with the 2015 launch of the EU Digital Single Market (DSM) strategy under the leadership of Commission President Jean-Claude Juncker. It then chased an international dimension with the pursuit of "digital sovereignty" under the Ursula von der Leyen presidency. Consequently, EU digital policies have emerged as a battleground where various stakeholders, including US tech giants, Member States, EU institutions, and interest groups, seek for influence and control.

In this special issue, we aim to unpack the multifaceted domain of "EU digital policies and politics" and the new challenges they pose to European regulators. To define EU digital policies and politics, we depart from a conventional approach that primarily examines the online behaviours of political actors (for example Koc-Michalska and Lilleker, 2017), embracing a perspective that analyses the dynamics between actors and policies regulating the digital realm.

The digital realm encompasses the technologies related to the non-analogue world, starting with the Internet and all the material (ie. servers, submarine cables, satellites, chips, etc.) and immaterial (ie. data) structures linked to it.[3] In other words, we refer to the digital realm as the assemblage of "technologies, infrastructures, data, and content based on and/or using electronic computing techniques" (Peters, 2016, in Couture and Toupin, 2019). The deployment of those digital technologies poses a range of new problems for contemporary European societies. For example, it generates new markets to be regulated, as a result of the exchange of non-exclusive and infinitely reproducible immaterial goods (Courmont and Galimberti, 2018). It also fuels new practices of political advertising, and facilitates the spread of illegal content online, to give other examples. The borderlessness of some of the practices, spaces and markets that digital technologies generate constitutes a direct challenge to the territoriality of states and their regulation. To sum up, as Büthe et al. (2022) put it in their analysis on artificial intelligence (AI), digital technologies raise new issues for states, as well as "familiar governance challenges in arguably novel ways". We define regulation as the various public policy instruments (Lascoumes and Le Galès, 2005) of public authorities to intervene in digital markets, encompassing a spectrum of approaches, from traditional hard law

Governance Act (DGA), Data Act and European Chips Act, and the forthcoming Artificial Intelligence Act (AIA).

3 For a thorough analysis of the "digital" artefacts, see: Kaufman and Jeandesboz, 2017.

regulations like the General Data Protection Regulation (GDPR)[4] to self-regulation by online platforms.

Digital regulation is relatively new in the landscape of European public policy. The interest of its study lies in the broad range of issues it addresses, including the impact of network effects on digital markets, issues surrounding the commercialization of private data and its implications for human rights, questions about online freedom of expression, and content moderation, the safeguarding of sensitive data and cybersecurity, as well as ensuring access to digital public services for citizens. With these concerns in mind, we are prompted to question what sets the EU's approach to this emerging realm of regulation apart from, on the one hand, other realms of regulation and, on the other hand, the approaches of other public regulators.

In recent EU digital regulations such as the GDPR or the DSA, different stakeholders (from decision-makers to digital rights advocates) have framed the challenges posed by the digital realm as questioning a set of fundamental principles and values, such as safeguarding private life, ensuring free access to information, protecting freedom of expression, enhancing security, and more. The scholarship on law and technology has developed the concept of "digital constitutionalism" to describe "the set of values and ideals that permeate, guide, and inform [...] normative counteractions against the alterations of the constitutional equilibrium produced by digital technology" (Celeste, 2019). This concept, rooted in legal theory, provides a valuable framework for our understanding of digital politics and policies. Indeed, it shows how digital regulation can be driven by competition between different fundamental principles and values. For example, in the GDPR case, the opposition between the protection of private life and the safeguard of a free market shaped most of the debates.[5] Thus, the literature on European digital constitutionalism (De Gregorio, 2022) helps us to unravel some of the key tensions at stake within digital regulation.

Drawing on this perspective as a starting point, this special issue addresses the following questions: What characterises and drives the EU's approach to digital regulation? How do EU policymakers try to answer the challenges

4 On the distinction between hard and soft law see, for example, Abbott and Snidal, 2000.

5 As a telling example of such tensions, the official title of the GDPR reads as follows: "Regulation on the protection of natural persons with regard to the processing of personal data and on the free movement of such data".

pertaining to specific characteristics of digital regulation? Who takes part in such regulation? As we delve into the various facets of EU digital policies, we aim to decipher the foundational principles and values that underpin the EU's regulatory stance in the digital sphere. In doing so, we strive to shed light on the path that the EU is charting in its pursuit of a "European digital constitution" (Bradford, 2023b).

The contribution of a political science perspective goes beyond merely identifying values. It extends to examining the "how" and the "who" in the realm of digital regulation. In this introduction, we will scrutinise the modes of regulation employed and the actors involved in shaping these regulations. By considering together the principles and values, the instruments of public policy and key stakeholders, we seek to provide a comprehensive understanding of the multifaceted landscape of EU digital politics. In doing so, we aim to contribute to a nuanced and insightful analysis of the EU's role in the ever-evolving digital realm.

Before delving into the five articles comprising this special issue and elucidating their contributions to a research agenda aimed at deciphering the central tensions within EU digital regulation, we will expound upon their defining and distinctive characteristics. We start by revisiting the origins of such policies, shedding light on the trajectory and controversies that have shaped EU digital regulation. Subsequently, we address the intriguing combination of fragmentation and transversality of EU digital regulation. Finally, we highlight how EU digital regulation serves as a microcosm encapsulating tensions between the national and global domains, thus necessitating further exploration.

## EU Digital Policies in the Making: The Reconfiguration of "Digital" Controversies Over Time

In addition to the vast scholarship on EU technology policy (Peterson and Sharp, 1998), this special issue contributes to an emerging body of literature investigating how EU legislations regulate the Internet and shape the governance of digital technologies (Schafer et al., 2020; Goyal et al., 2021).

This literature highlights in particular the various shifts that have marked the development of the digital policy domain, starting with the relative indifference of the European policy level for the development of the Internet (Christou and Simpson, 2007), until the 1990s and the sudden interest sparked by the commercialization of the Internet and the myriad of opportunities it created. This is well illustrated by the Bangemann report in 1994, which emphasised the newly assertive vision of the European Commission that "rather than remaining merely clients, we in Europe should consider following the evolution of the Internet closely, playing a more active role in the development of interlinkages" (European Commission, 1994).

The emergence of this interest in Internet policy came about in a broader context of deregulation of the telecom sector in the 1990s, and was perceived as a promising vector of growth for the European economy, up until the burst of the well-known dot-com bubble in the late 1990s. Gradually, the 2000s led to an acceleration of the European internal market approach applied to the digital. It was epitomised by the 2016 Digital Single Market Strategy, followed by the emergence of a more geopolitical and protectionist European agenda at the end of the last decade.

Aside from the trajectory and shifts of digital policies at the EU level, the literature on EU digital policy-making (Mărcuț, 2017; Newman, 2020) also looked at the complex political dynamics that underlie the resolution of "struggles" within such policies. It notably provided insights within the context of the growing influence of Big Tech companies in Brussels. A series of studies have investigated how data protection legislation was shaped at the European level (Kalyanpur and Newman, 2019; Schünemann and Windwehr, 2020; Christou and Rashid, 2021), underlining the contrasted success of interest groups and EU institutions in the adoption process of the EU GDPR for instance. The complex politics around the governance of digital platforms have also been unpacked from various perspectives as part of the recent EU copyright reform (Bonnamy, 2021, 2023a; Hermann, 2023) or in legislative processes related to data flows (Perarnaud, 2021) and cybersecurity (Timmers, 2018; Sivan-Sevilla, 2021).

While this scholarship provides for much needed observations of the conflicts and processes at the roots of specific EU legislations, it rarely focuses on the broader patterns and traits characterising EU digital policy efforts, and their evolutions over time.

## The Fragmentation and Transversality of EU Digital Policies

The special issue investigates what characterises and drives the EU's approach to digital regulation. Thus, one of its main features is a combination between fragmentation and transversality. Indeed, far from a uniform policy domain, EU digital policies have been marked by a gradual spillover towards a myriad of policy sectors (health, agriculture, environment, justice and home affairs...). Digital technologies and their providers have progressively set foot in these areas as a result of the so-called digital transformation of societies and economies, or by contrast, due to the immediate challenges that digital technologies pose for these policy fields.

Paraphrasing DeNardis' book - *The Internet in Everything* (2020) - showing how the diffusion of the Internet into the physical world radically escalates governance concerns and reshapes power structures between public and private actors, it appears that EU policies are similarly characterised by a trend towards embedding “digital into everything”. Digital issues are effectively mainstreamed into most, if not all, areas of EU competencies. What differs, however, from other types of mainstreaming, such as the gender one (Jacquot, 2010), is that it is not led by a single DG or a small set of actors with a clear mandate. In this sense, it can be compared to the mainstreaming of climate and environmental policies in the context of the Green New Deal for instance. This mainstreaming thus naturally generates streams of activities that do not necessarily align and can often become contradictory.[6]

As a result of these policy dynamics, and as digital technologies became pervasive across European countries, the European Commission has identified (or was invited to act in) many areas in which digital comes into play. For instance, the Directorate-General (DG) Migration and Home Affairs recognised the fight against terrorist online content as a crucial political priority, leading to the 2021 Terrorist Content Regulation. In the meantime, DG Employment, Social Affairs and Inclusion advocated in favour of addressing the concerning working conditions of gig workers of digital platforms, through a proposed directive in 2021, precisely at a time when DG Trade was busy refining what would become

6 The work of Bellon (2023) on the history and making of French digital policies over past decades illustrates a similar pattern. Bellon shows in particular how competing appropriation strategies can be at play between national politicians and civil servants, and how the growing prominence of digital issues exacerbated competition for ownership, and even sectoral capture, in the national governmental arena. These power struggles directly shaped the merging process of digital mainstreaming at the national level.

the newly adopted European Chips Act, another flagship legislation designed to strengthen the EU's semiconductor ecosystem. These are among the many recent legislative efforts spearheaded by the Commission, thus underlining that, instead of one clear actor setting the agenda, many entities are actually shaping the making process and implementation of EU digital policies. This institutional fragmentation is further exacerbated by individual rivalries and competition between EU leading figures, such as commissioners Margrethe Vestager (Competition), Thierry Breton (Internal Market) or Ylva Johansson (Home Affairs), who have attempted to steer the EU digital agenda in different directions between 2019 and 2023.

EU digital policies need thus to be understood in their multiplicity. The very term "EU digital policies" hides the fragmented nature of the myriad of parallel policy discussions and governance arenas constituting these policies. If arguably fragmented, these different and often unrelated policy streams can, however, be at times influenced or shaped by transversal patterns, such as the well-documented rise of the EU digital sovereignty discourse (Roberts et al., 2021). Disregarding this diversity in the power arrangements and loci in which these policies are defined and implemented is only helpful if one is to study such policies as a silo. However, as illustrated in this special issue by the broad scope of actors, processes and types of public action, EU digital policies are far more than a new policy silo. Most notably, they provide a stimulating ground for comparative research within and across policy sectors, for instance, to explore the evolving relationship between the EU, its Member States and the private sector in embedding the "digital" into European policies. That is why we eventually argue that one should conceptualise EU digital policies as a transversal category of public action. Moreover, comparative studies could be done with other policy areas also characterised by transversality, such as environmental policies.

## Sovereignty, Geopolitics and Big Tech: a Microcosm of Tensions between the National and Global

The state of fragmentation of EU digital policies leads to contradictions and inconsistencies between its various dimensions, but also between these policies and the rest of the EU policy acquis.

A case in point is the growing reflection of the digital sovereignty narrative in EU discourses. Though confusing (Lambach and Oppermann, 2022), the concept of digital sovereignty refers to the idea that States should reaffirm their authority over the Internet and protect their self-determination in the digital sphere (Musiani, 2022).

Since the early 2020s, noting the absence of a European equivalent to the US GAFA or the Chinese BATX,[7] EU leaders have been increasingly advocating for 'European digital sovereignty'. Recent literature has thoroughly explored this trend and its main drivers (Pohle and Thiel, 2020; Roberts et al., 2021; Bellanova *et al.*, 2022), with a particular focus on the recent EU initiatives to localise data within the European continent and adopt norms and technical standards to challenge the concentration of global technological market power in the United States and China. In this sense, EU digital policies can be viewed as a microcosm of the tensions structuring Internet governance more broadly at the global level (Schafer et al., 2020).

These tensions are of great significance in light of the implications of EU digital policies for actors far beyond the EU. Bradford (2020) proposed the concept of "Brussels effect" to characterise the dynamics at play. In her analysis of the global impact of the GDPR, she indeed illustrates the EU's capacity to export its values through its regulatory framework in response to the market power of external actors, particularly in the realm of new technologies. In this sense, Bradford argues that there would be three main models of regulation. She thus distinguishes the law-driven European model from the state-centric Chinese model and the US market-driven approach (Bradford, 2023a).

Development of the EU digital regulation under the von der Leyen Commission (2019-2024) invites us to nuance Bradford's analysis. Indeed, the growing "geopoliticisation" of EU digital policies (Broeders, Cristiano and Kaminska, 2023) has led the Commission to progressively deviate from its once market-centric and digitally-enthusiast regulatory approach. This policy evolution, fuelled by the digital sovereignty agenda, creates apparent tensions with the pre-existing EU digital policy acquis. This can be seen, for instance, in the field of Internet standards and standard-setting (Perarnaud and Rossi, 2023), where the EU has, over recent years, started normalising interventions on critical resources of the Internet architecture in stark contrast with its past policy approach in this area, condemning similar attempts. This tension is to be observed not only in this narrow domain but throughout the vast field of

7 The acronyms "GAFA" and "BATX" refer to the four largest North American and Chinese digital companies, Google, Apple, Facebook and Amazon, and Baidu, Alibaba, Tencent, and Xiaom respectively.

policies encompassing a digital component, as digital sovereignty discourses become gradually "operationalised" by the EU (Barrinha and Christou, 2022). These processes vividly illustrate the divergences between the economic and political interests of European actors and the norms and values championed by the EU and its Member States, as also evident in the debates on Internet governance (Claessen, 2020).

As underlined by the next section, the articles composing this special issue further investigate and unpack this European way of regulating the "digital".

## Presentation of the Articles

Firstly, Orsolya Gulyas proposes an Arendtian reading of contemporary EU digital policies. She unpacks the most recent EU data policies to expose the conceptual link between 'sovereignty' and 'competitiveness' in the European Union's digital agenda. Using the political theory of Hannah Arendt, she concludes that EU data policies are still profoundly rooted in modern frameworks of necessity and instrumentality rather than freedom and plurality. This is, for instance, illustrated by current imaginaries around the notion of digital sovereignty at the EU level, still prioritising economic thinking over human needs. Investigating the connection between EU digital policies and its ambitious environmental commitments, she shows that, as a result, current incompatibilities and power imaginaries prevent the meaningful twinning of the green and digital transitions. This piece thus helps us to grasp the principles and values at stake within EU digital policies and politics. It demonstrates the centrality of an economic perspective where values of competitiveness and free market prevail.

Julien Rossi's study of European data protection policy makes a significant contribution by uncovering that digital politics and policies do not consistently lag behind the technological developments they seek to regulate. Combining the analysis of archives from groups of interest, the EU, OECD and Council of Europe institutions, and interviews with policy-makers, he draws attention to the fact that a coalition of actors has maintained a classical (political) liberal definition of data protection since the 1970s to the adoption of the GDPR. Thus, the development of law doesn't necessarily follow one of the new technologies but it can be the other way around, or both (law and technology) can co-construct each other. Moreover, it confirms that EU digital public policies seem to escape neo-liberalization.[8] As such, Julien Rossi offers a nuanced perspective on the values underpinning the EU data protection policy. His long-term analysis shows that the right to privacy as a liberal value did prevail over the values of the free market in this specific and pioneer area of EU digital policies. Thus, in light of the other articles of this special issue, it underscores that the fragmentation of such policies and politics is also reflected in the values and principles at stake.

Examining e-health policy in Ireland and Austria, Chloé Bérut shows the strategic use of EU digital public policy instruments to promote a specific agenda. She demonstrates the centrality of *policy entrepreneurs* in implementing e-health policy in the two Member States under scrutiny. The European Commission 2004 eHealth action plan, a soft law, produced effects highly depending on national contexts. Bérut reveals that the EU guidelines were strategically mobilised by specific actors when broader political interest in eHealth was low. Not only does her piece underline the transversality of digital as a category of public action impacting national health policies, but it also highlights the crucial role of domestic actors in developing digital policies that may not be directly linked to EU initiatives but eventually result in the Europeanization of such policy.

Samuel Cipers and Trisha Meyer provide a thorough analysis of self-regulation practices of four of the biggest social media platforms (Google, Meta, TikTok and Twitter, now X). They specifically look at how they deal with political campaigning and compare this with the European Commission's expectations set out in a Code of Practice on Disinformation, first published in 2018, and a proposal for Regulation on Political Advertising. This piece contributes to understanding private actors' role in digital regulation. It clearly emphasises the centrality of self-regulation when it comes to speech regulation, implying potential broader consequences on vote behaviours. One of the paper's main contributions is to show how self-regulation leads to fragmentation within political advertising policy. Cipers and Meyer demonstrate that while having a critical intermediary role between public regulators and citizens, the digital platforms under study have very different definitions of "harmful content" and implement very different rules regarding political advertising, from total ban to controlled authorisation. Moreover, the paper mentions another layer of fragmentation between territories and types of political events. For instance,

8 Understood as economic, see also: Jabko, 2009; Woll, 2010.

they note a difference in treatment between electoral contexts, with the US presidential elections being much more monitored than others.

Finally, in his contribution to this special issue, Sebastian Heidebrecht develops the concept of *platform accountability* as an analytical tool to unpack recent EU legislation aimed at curbing the power of Big Tech companies. The paper shows how the Commission has aimed to hold platform companies to account and assesses the different qualities of the platform accountability framework that has emerged in the EU. Investigating the EU GDPR, Digital Market Act and DSA, on the basis of an innovative typology, he identifies the strengths of recent EU digital policies, notably in terms of democratic accountability, but also points out important challenges for the EU related to coherence, enforcement and credibility. This piece contributes to unpacking EU digital politics and policies by demonstrating how hard regulation plays as an instrument for the European Commission to regain power over private platforms. Furthermore, looking at accountability from the perspective of "democratic" and "constitutional" dimensions, it stresses the centrality of preservation of fundamental rights on the one hand and of market competition on the other hand, as the main drivers of the Commission's digital regulation.

## A Research Agenda to Explore Key Tensions in EU Digital Politics

The five articles included significantly contribute to the research agenda focused on defining the EU's approach to digital regulation. They achieve this by offering insights into key tensions within EU digital politics through two distinct lenses. Firstly, these papers illuminate the diversity of actors engaged in EU digital policies, showcasing their varying interests and approaches. Secondly, they provide an in-depth examination of the evolution of EU digital regulation by specifically delving into two multifaceted processes: Europeanization and neoliberalisation.

### Unravelling the Web of Actors in EU Digital Policies

Two central tensions emerge within the diverse contributions of this special issue regarding actors in EU digital policies and politics. The first is familiar in European studies, revolving around the classic struggle between Member States and EU institutions. National and European digital agendas can clash and diverge, particularly when the digital realm provides a means for EU institutions to venture into areas beyond their formal competencies, such as health or culture. The second tension, classical in a political economy perspective, is rooted in the contest for regulatory authority between public and private actors. As exemplified by Meta or Twitter/X, private companies tend to opt for distinct forms of self-regulation on a global scale. In contrast, the European Commission creates robust legal frameworks that apply to the EU single market. Moreover, regulating social platforms can directly impact political and social systems, influencing issues like freedom of expression and the management of private data, such as in the realm of health. Hence, the overarching question revolves around the locus of regulatory power and the ultimate determination of sovereignty. In the intricate landscape of EU digital politics, these tensions underscore the need to navigate a delicate balance between public and private interests while defining the contours of digital governance and sovereignty.

A complex and diverse cast of actors plays pivotal roles in EU digital policies and politics. While the institutional landscape often appears fragmented, resulting in varying levels of actor involvement based on the context, there is a remarkable coherence within the technology sector. Corporate Europe Observatory (2021) highlights this sector's dominance in lobbying expenditures at the EU level, particularly by major technology companies, raising pertinent questions about the influence of 'Big Tech' players on EU policy-making.

As such, our research agenda dives deep into understanding the roles and strategies of EU legislative actors and interest groups within the intricate landscape of EU digital politics. This exploration is critical for comprehending the competing interests of actors in EU digital policies (Perarnaud, 2021). Moreover, the contributions in this special issue provide an in-depth perspective on policy-making from the vantage point of these complex actors.

For example, we scrutinise the intricate relationship between the Internet platforms' industry, EU institutions, and Member States. Trisha Meyer and Samuel Cipers' article delves into the position of digital platforms within this complicated policy-making process, focusing on social media platforms and self-regulation. This work reveals the embroiled and often distant relationship between private digital actors and public legislators. Sebastian Heidebrecht's piece complements this perspective by shifting focus to the Commission's role and its approach to designing accountability mechanisms for these platforms. These two articles offer unique insights into the legitimacy of public action within the digital domain and the competition among diverse actors seeking to assert their authority as public or private Internet regulators. In the process, they contribute indirectly to the broader discourse surrounding "digital sovereignty", while also shedding light on its operational intricacies, as demonstrated by Orsolya Guylas' piece. Chloé Bérut's contribution provides a closer look at the role of Member States and how domestic actors actively participate in shaping integrated digital initiatives within health policies, driven by their unique interests.

## Tracing the Evolution of EU Digital Policies: Nuances of Europeanization and Neoliberalisation

In the preceding sections, we emphasised the absence of formal EU competence in digital policy, underscoring the imperative to unravel the historical development and regulatory paradigms shaping EU digital policies. This special issue grapples with this exigency with a dual emphasis: firstly, on the processes of Europeanization (Featherstone and Radaelli, 2003) and secondly, on the patterns of neoliberalisation within digital regulation.

Julien Rossi's contribution delves into the top-down Europeanization of data protection law, highlighting its surprising resilience since the 1970s, despite disruptive technological advancements. Complementing this perspective, Chloé Berut's piece examines 'strategic Europeanization' from a bottom-up approach, shedding light on the emergence of e-health policies within EU member states. Both articles underscore the intricate interplay of hard and soft law instruments as integral components of EU digital politics. Analysed as varying forms of Europeanization, these distinct instances of policy development illustrate the EU's capacity to foster diverse digital policies through multiple channels, even in the absence of formal competencies in the digital domain. These findings illuminate the multifaceted nature of EU digital politics, further elucidating the fragmentation previously discussed.

Using different theoretical frameworks—specifically, a cognitive approach to public policies and discourse analysis—Julien Rossi's and Orsolya Gulyas's respective research sheds light on the hues of neoliberalisation within EU digital policies and provides a nuanced perspective. Orsolya Gulyas underscores the centrality of an instrumental and economic perspective as the underlying logic behind discourses on sovereignty and freedom. On the other hand, Julien Rossi qualifies this perspective by demonstrating how the liberal political perspective prevails over an economic liberal one regarding data privacy. This aligns with recent scholarship, highlighting that the original and predominant "neoliberal rationality" (Dammann and Glasze, 2022) of these policies is now being challenged by a series of countermovements (Bonnamy, 2023b). This is evidenced by the rise of EU's digital sovereignty discourse and the "geo-dirigiste" turn of the Commission in its digital industry policy (Heidebrecht, 2023; Seidl and Schmitz, 2023), which are increasingly shaping the EU's approach to digital regulation.

## Conclusion

As elucidated in this introduction, digital regulation is a relatively recent policy priority for European regulators, characterised by a complex web of concerns and tensions surrounding fundamental values. These include the delicate balance between the protection of individual privacy and the promotion of a free market, as well as competing interpretations of freedom of expression.

The multifaceted nature of digital regulation makes it a challenging subject for the social sciences. Its transversal application across virtually all sectors of public policy, combined with its fragmented oversight by a diverse array of regulators, from specific Directorates-General within the European Commission to self-regulation by private companies, underscores its complexity. However, we contend that digital policies and politics serve as a microcosm of tensions not only among values but also across different levels of governance. This is exemplified in the ongoing struggles between the EU, the United States, and China over the regulation of major tech companies.

The papers in this special issue delve into these diverse tensions, offering two distinct entry points for understanding the EU's approach to digital regulation: the role of key actors and the underlying processes. Our contributors shed light on the various actors in the landscape of EU digital politics and policies, emphasising the pivotal roles played by the European Commission, national governments, and tech giants. With respect to the processes characterising the EU's digital approach, our papers underscore the significance and varied forms of Europeanisation processes, as well as the influence of neoliberalisation. A crucial contribution lies in the revelation that recent dynamics of EU digital regulation are shifting towards a focus on political liberties rather than merely economic considerations, while still being firmly rooted in a market-building perspective.

This special issue offers a comprehensive exploration of the complexities inherent in the EU's digital regulation policies, providing valuable insights into the intricate interplay of actors, processes, and values. Through the diverse and rich contributions of our authors, we aim to work towards a deeper understanding of the European approach to regulating the digital realm, which is of utmost relevance in our increasingly interconnected world.

## Bibliography

**Abbott, Kenneth and Snidal, Duncan** (2000), "Hard and Soft Law in International Governance", *International Organization*, vol. 54 (3), p. 421-456.

**Barrinha, André and Christou, George** (2022), "Speaking sovereignty: the EU in the Cyber Domain", *European Security*, vol. 31 (3), p. 356-376.

**Bellanova, Rocco, Carrapico, Helena and Duez, Denis** (2022), "Digital/sovereignty and European security integration: an introduction", *European security*, vol. 31 (3), p. 337-355.

**Bellon, Anne** (2023), *L'État et la Toile. Des politiques de l'internet à la numérisation de l'action publique*, Vulaines-sur-Seine, Éditions du Croquant.

**Bendiek, Annegret and Römer, Magnus** (2019), "Externalizing Europe: the global effects of European data protection", *Digital Policy, Regulation and Governance*, vol. 21 (1), p. 32-43.

**Bonnamy, Céleste** (2021), "Market, culture, and open access: European copyright and the renewal of a historical clash of values in the digital age", *in* François Foret et Jana Vargovcíková (eds), *Value Politics in the European Union. From Market to Culture and Back*, London, Routledge, p. 58-77.

**Bonnamy, Céleste** (2023a), "From liberalisation to regulation: managerial political work in the European digital copyright policy (2014-2019", *Journal of European Public Policy.*

**Bonnamy, Céleste** (2023b), *Avec la culture, contre internet ? Sociologie de la régulation européenne du droit d'auteur numérique (2004-2019)*, Thèse de doctorat, Université libre de Bruxelles et Université Paris 1 Panthéon-Sorbonne.
Bradford, Anu (2020), *The Brussels effect: How the European Union rules the world*, Oxford, Oxford University Press.

**Bradford, Anu** (2023a), *Digital Empires: The Global Battle to Regulate Technology*, Oxford, Oxford University Press.

**Bradford, Anu** (2023b), "Europe's Digital Constitution", *VA. J. INT'L L.*, vol. 64 (1).

**Broeders, Dennis, Cristiano, Fabio and Kaminska, Monica** (2023), "In Search of Digital Sovereignty and Strategic Autonomy: Normative Power Europe to the Test of Its Geopolitical Ambitions", *Journal of Common Market Studies*, vol. 61 (5), p. 1261-1280.

**Büthe, Tim, Djeffal, Christian, Lütge, Christoph, Maasen, Sabine and von Ingersleben-Seip, Nora** (2022), "Governing AI – attempting to herd cats? Introduction to the special issue on the Governance of Artificial Intelligence", *Journal of European Public Policy*, vol. 29 (11), p. 1721-1752.

**Celeste, Edoardo** (2019), "Digital constitutionalism: a new systematic theorisation", *International Review of Law, Computers & Technology*, vol. 33 (1), p. 76-99.

**Christou, George and Rashid, Imir** (2021), "Interest group lobbying in the European Union: privacy, data protection and the right to be forgotten", *Comparative European Politics*, vol. 19, p. 380-400.

**Christou, George and Simpson, Seamus** (2007), "Gaining a stake in global internet governance: the EU, ICANN and strategic norm manipulation", *European journal of communication*, vol. 22 (2), p. 147-164.

**Claessen, Eva** (2020), "Reshaping the internet–the impact of the securitisation of internet infrastructure on approaches to internet governance: the case of Russia and the EU", *Journal of cyber policy*, vol. 5 (1), p. 140-157.

**Corporate Europe Observatory** (2021), "The lobby network: Big Tech's web of influence in the EU", *Corporate Europe Observatory* [Blog post], 31 August, <https://corporateeurope.org/en/2021/08/lobby-network-big-techs-web-influence-eu>.

**Couture, Stephane and Toupin, Sophie** (2019), "What does the notion of "sovereignty" mean when referring to the digital?", *New Media & Society*, vol. 21 (10), p. 2305-2322.

**Dammann, Finn and Glasze, Georg** (2022), "Governing digital circulation: the quest for data control and sovereignty in Germany", *Territory, Politics, Governance*, vol. 11 (6), p. 1100-1120.

**De Gregorio, Giovanni** (2022), *Digital constitutionalism in Europe: Reframing rights and powers in the algorithmic society*, Cambridge, Cambridge University Press.

**DeNardis, Laura** (2020), *The Internet in everything*, New Haven, Yale University Press.

**European Commission** (1994), *Europe and the Global Information Society: Recommendations to the European Council*, ISPO, Brussels, 26 May, <www.ispo.cec.be/infosoc/backg/bangeman.html>.

**Featherstone, Kevin and Radaelli, Claudio M.** (eds) (2003), *The politics of Europeanization*, Oxford, Oxford University Press.

**Goyal, Nihit, Howlett, Michael and Taeihagh, Araz** (2021), "Why and how does the regulation of emerging technologies occur? Explaining the adoption of the EU General Data Protection Regulation using the multiple streams framework", *Regulation & Governance*, vol. 15 (4), p. 1020-1034.

**Heermann, Max** (2023), "Undermining lobbying coalitions: the interest group politics of EU copyright reform", *Journal of European Public Policy*, p. 1-29.

**Heidebrecht, Sebastian** (2023), 'From Market Liberalism to Public Intervention: Digital Sovereignty and Changing European Union Digital Single Market Governance', *Journal of Common Market Studies.*

**Jabko, Nicolas** (2009), *L'Europe par le marché: Histoire d'une stratégie improbable*, Paris, Presses de Sciences Po.

**Jacquot, Sophie** (2010), "The paradox of gender mainstreaming: Unanticipated effects of new modes of governance in the gender equality domain", *West European Politics*, vol. 33 (1), p. 118-135.

**Kalyanpur, Nikhil and Newman, Abraham L.** (2019), "The MNC-Coalition Paradox: Issue Salience, Foreign Firms and the General Data Protection Regulation", *Journal of Common Market Studies*, vol. 57 (3), p. 448-467.

**Kaufmann, Mareile and Jeandesboz, Julien** (2017), "Politics and 'the digital': From singularity to specificity", *European Journal of Social Theory*, vol. 20 (3), p. 309-328.

**Koc-Michalska, Karolina and Lilleker, Darren** (2017), "Digital politics: Mobilisation, engagement, and participation", *Political Communication*, vol. 34 (1), p. 1-5.

**Lambach, Daniel and Oppermann, Kai** (2022), "Narratives of digital sovereignty in German political discourse", *Governance*, vol. 36 (3), p. 693-709.

**Lascoumes, Pierre and Le Galès, Patrick** (2005), *Gouverner par les instruments*, Paris, Presses de Sciences Po.

**Mărcuț, Mirela** (2017), *Crystalizing the EU digital policy: an exploration into the digital single market*, Cham, Springer.

**Musiani, Francesca** (2022), "*Infrastructuring* digital sovereignty: a research agenda for an infrastructure-based sociology of digital self-determination practices", *Information, communication & society*, vol. 25 (6), p. 785-800.

**Newman, Abraham L.** (2020), "Digital Policy-Making in the European Union: Building the New Economy of an Information Society", *in* Helen Wallace, Mark A. Pollack, Christilla Roederer-Rynning and Aladair R. Young (eds), *Policy-Making in the European Union*, Oxford, Oxford University Press, p. 275-296.

**Perarnaud, Clément** (2021), "A step back to look ahead: Mapping coalitions on data flows and platform regulation in the Council of the EU (2016-2019)", *Internet Policy Review*, vol. 10 (2), p. 1-21.

**Perarnaud, Clément and Rossi, Julien** (2023), "The EU and Internet standards: beyond the spin, a strategic turn?", *Journal of European Public Policy.*

**Peters, Benjamin** (2016), "Digital", *in* Peters Benjamin (ed.), *Digital Keywords: A Vocabulary of Information Society and Culture*, Princeton, Princeton University Press, p. 93-108.

**Peterson, John and Sharp, Margaret** (1998), *Technology policy in the European Union*, Basingstoke, Macmillan.

**Pohle, Julia and Thiel, Thorsten** (2020), "Digital sovereignty", *Internet Policy Review*, vol. 9 (4), p. 1-19.

**Roberts, Huw *et al.*** (2021), "Safeguarding European values with digital sovereignty: An analysis of statements and policies", *Internet Policy Review*, vol. 10 (3), p. 1-28.

**Schafer, Valérie *et al.*** (2020), "The internet and the European market from a multidisciplinary perspective: a "round-doc" discussion", *Internet Histories*, vol. 4 (2), p. 217-227.

**Schünemann, Wolf Jürgen and Windwehr, Jana** (2021), "Towards a 'gold standard for the world'? The European General Data Protection Regulation between supranational and national norm entrepreneurship", *Journal of European Integration*, vol. 43 (7), p. 859-874.

**Seidl, Timo and Schmitz, Luuk** (2023), "Moving on to not fall behind? Technological sovereignty and the 'geo-dirigiste' turn in EU industrial policy", *Journal of European Public Policy.*

**Sivan-Sevilla, Ido** (2023), "Supranational security states for national security problems: governing by rules & capacities in tech-driven security spaces", *Journal of European Public Policy*, vol. 30 (7), p. 1353-1378.

**Timmers, Paul** (2018), "The European Union's cybersecurity industrial policy", *Journal of Cyber Policy*, vol. 3 (3), p. 363-384.

**Woll, Cornelia** (2010), "L'Union européenne : une machine à libéraliser ?", *Politique européenne*, vol. 31 (2), p. 215-220.

**Céleste Bonnamy** is a postdoctoral researcher in the Department of Political Science at LUISS Guido Carli in Rome. She holds a PhD in Political Science from the Université libre de Bruxelles and Université Paris 1 Panthéon-Sorbonne. Her current research focuses on the role of cultural industries in digital regulation and policy-making.

celeste.bonnamy@ulb.be

**Clément Perarnaud** is a postdoctoral researcher at the Vrije Universiteit Brussel (VUB). He is the secretary of the Global Internet Governance Academic Network (GigaNet) and member of the working group on Internet governance and regulation of the Center Internet et Société (CIS-CNRS). His most recent research work focuses on the making of EU digital policies and standards, as well as negotiations and power plays in Internet governance.

clement.perarnaud@vub.be

**POLITIQUE EUROPÉENNE**
**N° 81 | 2023**
Orsolya Gulyás
p. 28-53

**Digital Sovereignty, Competitiveness, and the Illusion of Freedom**

**An Arendtian Approach to EU Digital Policy**

The article exposes the conceptual link between 'sovereignty' and 'competitiveness' in the European Union's digital strategy and policy by examining its most recent data policies. It assesses to what extent they represent a conceptual shift in digital policymaking towards a more horizontal understanding of power in line with the EU's newly proposed human-centered approach and the green transition. Using the political theory of Hannah Arendt, it concludes that EU data policies are still profoundly rooted in modern frameworks of necessity and instrumentality rather than freedom and plurality, and suggests that this prevents the meaningful twinning of the green and digital transitions.

**Souveraineté numérique, compétitivité, et l'illusion de la liberté**

**Une approche arendtienne de la politique numérique de l'UE**

Cet article expose le lien conceptuel entre « souveraineté » et « compétitivité » dans la stratégie et la politique numériques de l'UE en examinant ses politiques les plus récentes en matière de données. Il évalue dans quelle mesure elles représentent un changement conceptuel dans l'élaboration des politiques numériques, vers une approche plus horizontale du pouvoir, conformément à la nouvelle approche centrée sur l'humain proposée par l'UE et à la transition verte. En utilisant la théorie politique de Hannah Arendt, l'article fait voir que les politiques de l'UE en matière de données sont encore profondément enracinées dans des cadres modernes de nécessité et d'instrumentalité plutôt que de liberté et de pluralité, et suggère que cela empêche le couplage significatif des transitions verte et numérique.

# Digital Sovereignty, Competitiveness, and the Illusion of Freedom

## An Arendtian Approach to EU Digital Policy

**Orsolya Gulyás**
Brussels School of Governance, Vrije Universiteit Brussel, Belgium

## Introduction

In recent years, the development of the digital sphere has become increasingly tied to economic and foreign policymaking. Technological development and regulation have been taking on a more and more prominent role in geopolitical debates. This is not surprising given the widespread changes that the digital transition spurs in our daily lives as various concerns arise related to issues such as privacy, surveillance, disinformation, cybersecurity, antitrust violations, or trade disputes, among others. One recent tension is the emergence of a drive for sovereignty in digital politics and policymaking. Digital sovereignty has become a central notion in the European Union's (EU) digital strategy, often invoking imaginaries of being constrained and dominated by other actors.

This article aims to foreground the economic dimension of EU digital policy-making and its role in the sovereigntist turn. While this has been addressed in some of the scholarly literature on digital sovereignty, (e.g. Bellanova *et al.*, 2022; Pohle and Thiel, 2020; Thumfart, 2022), it has not been sufficiently linked to neoliberal policymaking on a conceptual level. Other literature points to a recent shift from neoliberal to more neo-mercantilist approaches to EU digital policy (e.g. Farrand and Carrapico, 2022; Schmitz and Seidl, 2023), however, these tend to focus on novel policy tools and do not sufficiently highlight the continuity of the neoliberal logic that keeps underlying EU policymaking. As such, the article builds on the tracing and analysis of the 'competitiveness' discourse in the recent data strategy and data policies of the EU. The analysis reveals the logic of competitiveness as central to digital policymaking and politics and argues that the recent emergence of the sovereignty discourse is not a novelty, rather a continuation of the EU's 'competitiveness, growth and jobs' agenda that was put in motion with the Lisbon strategy, setting the scene for the EU's long-standing pursuit of neoliberal policymaking. It suggests treating competitiveness as a discourse of power that perpetuates a fundamentally modern conceptualization of agents and relations, and which is in tension with the EU's newly proposed, more human-centered policymaking and the twinning of the green and digital transitions.

The paper introduces a novel approach to the analysis of the EU digital sovereignty discourse through the lens of Hannah Arendt's political thought, shedding light on the conceptual link between digital sovereignty and economic competitiveness. It shows that their common root stems from the conceptualization of politics that regards power as an instrumental force by which those 'on top' can act alone and impose their will upon others. This implies a vertical imaginary of power relations that foregoes the Arendtian conceptualization of politics as the freedom experienced when acting together with plural equals. As such, the sovereignty discourse indicates a desire for independence and control, and simultaneously a solution to cope with vulnerability that results from the interdependence that is inherent in the digital realm.

Indeed, the research argues that the EU's digital sovereignty discourse and approach to digital policymaking are in fact rooted in tropes of necessity and instrumentality, rather than the EU's capacity to act freely and/or independently. The anxiety about losing economic competitiveness underlies recent EU digital policymaking to a much larger extent than the EU's capacity to act without constraints by others, resulting in a misconceptualization of sovereignty as freedom.

Tracing this conceptualization also reveals a fundamental contradiction in the EU's recent twinning of the digital and green transitions. Namely, if the gist of the EU's digital politics lies in a desire to control and dominate others in the pursuit of the economic growth and competitiveness of EU businesses, then this contradicts the essence of the green transition, which would require that policymaking takes seriously the limits to economic growth. This would inevitably result in a shift in mindset towards less competitive and more co-operative modes of conceiving of politics.

Given that both the digital and green transitions represent central challenges for 21st century politics and policymaking, and the fact that both are inherently global issues requiring meaningful global co-operation, examining the underlying assumptions and power imaginaries surrounding them is of paramount importance. By emphasizing the role of language and discourse as an active shaper of policy meanings, the research contributes to the emerging field of critical digital policy studies.

The article first gives an overview of the academic literature on (EU) digital sovereignty, then introduces the Arendtian theoretical framework that follows on from the work of Nicole Dewandre (2018), which provides the core innovation in this paper. The following chapter describes the methodology and case study, consisting of an analysis of the European Union's recently proposed data strategy and policies. In the discussion of the case study, the paper examines the implications of the theoretical framework and offers a novel lens through which to understand European digital policymaking in the 21stcentury. The paper concludes with suggestions on how to step out of the modern framework in European digital policymaking by making room for the notions of satiety and vulnerability, which could ease the current tension underlying the twinning of the green and digital transitions, and help make the most of their coupling.

## Digital Sovereignty in the European Union

The internet has for long been thought of as the communication network that spans the globe and connects its diverse communities (Castells, 1996). This connectedness, however, has also more recently brought to the surface various political actors' profound anxieties about losing control and becoming dependent on others (see e.g. Pohle and Thiel, 2020; Thumfart, 2022; Obendiek, 2021). In a speech given in 2021, Thierry Breton, the EU's Commissioner for the Internal Market talked about digital sovereignty and the geopolitics of technology, stating that Europe *"remains open, but on its own terms. It makes its own choices and draws up its own rules, and is not afraid of imposing them on its partners. We have entered a global race in which the mastery of technologies is central."* While there is no one single, coherent understanding within the EU and its institutions about the exact meaning of European digital sovereignty, this formulation by the Commissioner in charge of the internal market (and digital policy) is illuminating because it invokes 1) the EU's desire to act freely and independently of other actors, and 2) its desire to dominate others by imposing its will and by winning the global race that characterizes technological development.

Discourses about digital sovereignty started to surface first in China and Russia as a response to their perception of US hegemony in the digital world (Creemers, 2020; Thumfart, 2022). Claims to sovereignty in the digital domain by such state actors are remarkable but coherent with their political stances and ambitions; however, in the case of the European Union it is more puzzling. As a political and economic union of Member States founded on the very idea that their interdependence would be the guarantor of peace on the continent, the EU is not new to the idea of cooperating with others. Indeed, placing digital sovereignty at the center of EU digital politics and policies positions the EU's actions always in relation to an outside, external dimension that risks the preclusion of much-needed global digital co-operation (Obendiek, 2021, 2).

The notion of digital sovereignty has gained renewed attention in academic literature after the 2013 Snowden revelations about the global governmental surveillance programs run by the United States and its allies (see e.g. Floridi, 2020; Mueller, 2017; Pohle and Thiel, 2020; Thumfart, 2022). In general, the term has been identified with the tendency of diverse actors (state, corporate or individual/civil society) to strive for independence and self-determination in the digital sphere (Couture and Toupin, 2019). The literature has distinguished the state, economic and individual dimensions of digital sovereignty and has acknowledged the intertwining of these perspectives (Pohle and Thiel, 2020). The state dimension is typically associated with the security of digital infrastructures. The economic dimension often relates to the perceived market dominance of mostly US, but also Chinese technology companies and emphasizes the need for supporting innovation with the aim that domestic companies become better positioned to compete with foreign ones in the global market. The individual dimension highlights the need for more user autonomy and looks beyond state-centered approaches, focusing on the ability of citizens to have more control when it comes to digital infrastructures and communication (Pohle and Thiel, 2020).

Sovereignty is traditionally a central concept in politics, indicating the superior authority of the state to exercise power, and importantly, it is bound to a given territory. As such, this notion is often perceived to be in contradiction with the essence of the internet as a networked technology. Moreover, it has been argued that the inherently territorially-bounded understanding of digital sovereignty is conceptually incompatible with the internet as a global communication network (Pohle and Thiel, 2020; Shahin, 2007). However, the concept is still widely used in politics and policymaking and scholarly contributions have typically highlighted the socio-economic and political pitfalls associated with the fragmentation of the internet, the risk of which is thought to be fueled by the pursuit of digital sovereignty (Mueller, 2017).

Other scholars have brought attention to the diminishing power of state actors in the digital space as transnational companies increasingly set the rules (van Dijck *et al.*, 2018) and tech giants have been described as possessing 'functional sovereignty' with regard to their activities (Pasquale, 2018).

When it comes to the EU, Rocco Bellanova *et al.* (2022) give an extensive overview of the puzzle that European sovereignty claims create, given that the EU is not a nation state. They distinguish scenarios of sovereignty *over* the digital, when the EU aims to develop and control digital infrastructures, and sovereignty *through* the digital, which indicates the use of digital tools for European (security) governance. The article also points to the concept's origins on the member state level: in Europe the first instances of the notion appeared in France in the mid-2000s over issues of data privacy and economic competitiveness, followed by Germany where concerns over security arose in the wake of the Snowden revelations. These two member states then brought the issue to the European political scene, claiming that it was better dealt with via the working together of member states at the EU level, where digital and technological development was long seen as the driver of economic growth (see Gulyás and Shahin, 2023; Mansell, 2014; Schmitz and Seidl, 2023).

Indeed, in 2016 Günther Oettinger, the then EU Commissioner for the Digital Economy and Society, talked about the creation of a European digital single market as a way to ensure the EU's digital sovereignty and competitiveness (Bellanova *et al.*, 2022). By 2020, with the current von der Leyen Commission, digital sovereignty has become a central element of the EU's digital strategy and the digital transition has been made one of the main priorities of EU policymaking, launching Europe's 'Digital Decade' and building 'A Europe fit for the digital age' (European Commission, 2022a).

While the role of global economic competition and competitiveness has been acknowledged in works about the EU's digital sovereignty discourse, it has rarely been critically examined in the framework of neoliberal policymaking (see for example Roberts *et al.*, 2021). The work of Luuk Schmitz and Timo Seidl (2023) stands out in this regard, tracing the roots of the rise of Europe's open strategic autonomy and sovereignty discourse to embedded neoliberalism. While their work focuses on trade policy, they acknowledge that these issues have appeared in the EU's digital or industrial policy too. The authors point to a recent shift towards neo-mercantilist ideas as the discursive hegemony of the efficiency of 'the market' is fading amidst increasing geopolitical competition, prompting these new strategies to protect, transform and project the single market. In a similar vein, Heidebrecht (2023) highlights the EU's shift from a market-liberal to a more public-interventionist governance of the digital realm, and suggests that policy aims have gradually broadened to include further objectives next to economic competitiveness.

Benjamin Farrand and Helena Carrapico (2022) also point to the implications of the competitiveness agenda for digital sovereignty, highlighting its transformational role for the relationship between the public and private sectors in EU cybersecurity policy. Similarly to L. Schmitz and T. Seidl (2023), they argue that the EU has entered a new phase of regulatory mercantilism in its cybersecurity policy, where foreign businesses are treated as much of a threat as foreign powers, and EU businesses are conceived of as the guardians and embodiments of EU values (Farrand and Carrapico, 2022).

Taking on board the rich analyses offered in the abovementioned works, in this article I suggest that what we see with the rise of the digital sovereignty discourse is more a continuation of the old, entrenched ways of neoliberal policymaking, rather than a shift. Taking to heart the needs and desires of EU economic agents and conceiving of the internal market as the home base for European companies where they can grow in order to become fit to compete on the global stage is an approach that has been defining EU policymaking at least since the Lisbon strategy launched in 2000 (Dewandre and Gulyás, 2018). Reflecting on the changing geopolitical conditions later on in my discussion of the case study, I situate my thinking in the wake of William Davies (2014), pointing to the global economic race that underlies neoliberal policymaking in the EU and beyond, perpetuating the competitiveness agenda.

Most importantly however, the primary aim and contribution of this paper is a critical analysis of and reflection on the conceptual imaginaries of the types of power relations that are represented by the linking of the EU's digital sovereignty and economic competitiveness discourses. The next section introduces the Arendtian conceptual framework that underpins and inspires this analysis.

## Power and politics: key concepts in Hannah Arendt's thought

This study rests on an Arendtian approach to the analysis of EU policymaking as developed by Nicole Dewandre in her article *Political Agents as Relational Selves: Rethinking EU Politics and Policy-Making with Hannah Arendt* (2018). The applicability of this Arendtian framework to EU policies was first introduced in a paper offering a critical metaphor analysis of strategic EU policy documents (Dewandre and Gulyás, 2018). However, before introducing N. Dewandre's interpretation, I wish to highlight some main features of Hannah Arendt's thought and define the notions and concepts that are key to understanding the arguments put forth.

In her influential work, *The Human Condition* (first published in 1958), Hannah Arendt introduced a highly original understanding of politics and the place and realm it occupies in human affairs. Her account examines and challenges some of the Western tradition's most fundamental ideas and notions about politics, power, authority, political action, or freedom, amongst others. Tracing roots as far back as ancient Greek political philosophy, Arendt's analysis points out that our very ideas of politics, independent of being left or right-wing, already carry within them conceptualizations that are incompatible with meaningful participation, the acknowledgement of human diversity ('plurality' as she refers to it), as well as with the (democratic) promise of exchange of opinions and debate among equals (Villa, 2000). As such, her work is useful to shed light on what could constitute a way out of our entrenched ways of thinking that prevent politics and policies from generating meaning in today's EU (Dewandre, 2018).

The most prominent idea put forth in *The Human Condition* is the tripartite distinction of human activities into the realms of 'labor', 'work', and 'action'. Very briefly, labor means taking care of our biological needs (e.g. eating or sleeping) in order to survive. The realm of work designates human activities that are occupied with making or fabricating things, or shaping our environment, such as building a table. Finally, the third realm of human activities is action, which occurs directly between people. For this reason, according to Arendt, politics can only take place in the realm of action. The main feature of this category of human activities is the frailty and unpredictability that are inherent to acting in the world, as these actions always already take place in a web of relations where other actors are present too. This emphasis on the uncertainty and unpredictability of the outcome of one's actions may suggest that political action is necessarily a frustrating and unpleasant ordeal, forcing us to experience our vulnerability and lack of control. However, for Arendt this was quite to the contrary: she highlights that it is precisely this unpredictability and contingency that gives rise to the experience of one's freedom by initiating something new in the public realm. Indeed, for Arendt, freedom is constituted by the possibility to begin, which is denoted by her notion of 'natality'. Furthermore, it is in the realm of action that actors are able to access, reveal, and act out their own unique identities in public. Corresponding to this, a central notion in Arendt's theory of action is 'plurality', which is the acknowledgment of human diversity: we are all equal as humans but each of us is also unique (Villa, 2000).

In the Western tradition, this kind of plural political action is difficult to find. Instead, sovereign action - acting alone and upon others – is the preferred norm. Indeed, in today's world other political actors are routinely regarded as obstacles to one's freedom, and certainly not as sources of it. As Arendt shows, this is because of the conceptualization of politics as if it was akin to fabrication (work), where it is possible for humans to exert rational control upon an object, leading to the "traditional substitution of making for acting" (Arendt and Canovan, 1998, 220). This instrumentality that is found in traditional conceptualizations is, according to Arendt, anti-political with its refusal to accept human plurality and the uncertainty of the outcome of action. Importantly, she viewed political action as the opposite of violence, where power arises as a result of diverse equals acting together, instead of imposing their will on one another (Bernstein, 2011).

Following this very brief introduction to Hannah Arendt's rich and dense system of thought, the next section explains in more detail how it can be transposed and employed to analyse (EU) policymaking. Using the tripartite lens of labor, work, and action is a novel way to capture policy rationales and is a productive and revelatory tool to understand to what extent politics and policymaking are driven by ideas of necessity, instrumentality, and/or freedom.

## Agents and relations: an Arendtian approach to analysing EU policy

Drawing on Hannah Arendt's critique of modernity, Nicole Dewandre (2018) situates her account of EU policymaking in the emergence of the 'competitiveness, growth and jobs' agenda that indicated the rise of neoliberal policies in the mid-80s. During this time, the EU notably shifted its discourse from talking about a 'common' market and started to use the expression 'internal' market. This change of language also signaled a change of purpose for the then-European Community as the underlying rationale shifted from enhancing interdependence among member states to creating a home base for European industry to grow so that EU businesses become better equipped to compete on the global economic stage. This shift away from fostering interdependence puts into question the EU's original postnational and postmodern character as a political structure, a manifestation of which is promoting illusions of sovereignty (Dewandre, 2018).

Dewandre's approach (2018) operationalizes the Arendtian framework by analysing contemporary politics and policymaking along two strands: agents and relations. In her work she reinterprets Arendt's tripartite distinction of human activities (labor, work, action) into modes of relation.

In modernity, relations between two entities are generally conceptualized in terms of causality, relying on the asymmetry between the 'knower' and the 'known'. The knower, like a cause, is active, dominant, and in control. Meanwhile, the known – or the effect - is passive, dominated and under control. When it comes to agents, they are similarly conceptualized as 'rational subjects' who are merely interest-led beings and are either in control, or being controlled. This posits power relations on a vertical scale, with a distinction of those being on the top and in control, acting upon others who are on the bottom, resulting in the conceptualization of power as if it was a mechanical force. Indeed, it is the 'being on top' or 'in control' that gives a sense of power according to the modern imaginary, hence the prominence of a desire for domination. In N. Dewandre's interpretation this modern, vertical mode of relation corresponds to that of 'work' in Arendt's distinction of the three types of human activities. It is an essentially instrumentalizing relation that rests upon a means-to-end logic. It is in this type of activity that humans realize what they have envisaged, acting upon an object or on another as if they were an object, leading to the self being in control, and the object being under control (Dewandre, 2018).

On the other hand, the modern worldview conceptualizes horizontal relations among equal entities as being independent of each other, i.e. they are not bound by a causal relationship. They are thought of as being in balance, reaching an equilibrium of forces that are adversarial and mutually neutralizing. Hence, relations among equals are seldom characterized as productive in the modern imaginary: they remain impotent interactions where the parties experience the presence of the other as constraining their own freedom and limiting their capacity to act (Dewandre, 2018).

In distinction to such reductionist approaches to both vertical and horizontal relations, the Arendtian concept of 'action' is a type of activity rooted in plurality. This can be characterized as a horizontal mode of relation that is productive, as plurality for Arendt is centered on the fact that as humans, we are all equal but also unique. Humans can experience their freedom by mutually revealing and recognizing each other's identity via the interaction of equal but unique selves, accepting the unpredictability of the outcomes of their (inter)actions. According to the Arendtian framework, this is where (nonhegemonic) politics can take place: it is rooted in and strives for freedom, rather than control and domination, and power does not act as a force over others but springs up when people act together (*Ibid.*).

The third, surrounding mode of relation, corresponding to 'labor' as a type of activity in Arendt's thinking, designates the other neither as an object nor a self, but rather as a milieu that needs to be coped with. It is rooted neither in instrumentality (work) nor freedom (action), but in the notion of necessity as it has to do with survival and adaptation to the environment: the facts of life that cannot be escaped as they relate to keeping up our biological life processes (*Ibid.*).

Using Hannah Arendt's critique of modernity and this tripartite distinction of human activities (labor, work, action) rearticulated as modes of relation (necessity, instrumentality, and freedom, respectively) is useful for the analysis of EU policies for two main reasons: 1) it helps identify the roots of the thinking behind policy rationales, and 2) it can show the way out of entrenched patterns of thinking that hinder the development of policies that are fit for the 21st century, where the interdependence of political actors is increasingly a given.

## Competitiveness and the intertwining of public and private actors

The presence of the entrenchment of the modern worldview grounded in vertical power relations in EU policymaking has been confirmed by a study conducting a critical metaphor analysis of 30 years of high-level, strategic EU policy documents between 1985 and 2014 (Dewandre and Gulyás, 2018). One of the main findings was what the authors termed the *sensitive inversion*.

The study shows that in line with the modern worldview, EU policy strategies overwhelmingly treated human beings as if they were purely rational without desires and needs. Furthermore, humans were also reduced to functional beings in the policy strategies as they were overwhelmingly referred to as consumers, human resources, workforce, or were regarded as commodities that need to develop skills in order to satisfy business needs. At the same time, economic agents (business, industry, markets, the economy, etc.) were treated as sensitive personae with needs and desires, but were also expected to behave rationally as profit-maximizers. This *sensitive inversion* demonstrates that neoliberal policymaking essentially rendered policymakers to view the world through the eyes of economic agents, keeping their interests and desires at heart much more clearly and unequivocally than the needs of humans, all in the name of increasing competitiveness and perpetuating economic growth (Ibid.). Indeed, the EU has often been critiqued for its neoliberal policymaking, one of the main features of which is the personification of economic agents and the simultaneous debasement of people (e.g. Harvey, 2005; Holborow, 2015; Mirowski and Plehwe, 2009).

Describing neoliberalism as the *'pursuit of the disenchantment of politics by economics'*, William Davies (2014, 7) assigns central importance to 'competitiveness' as a defining element of neoliberal discourse, the performative power of which needs to be taken seriously since political action takes place on its basis. He draws attention to an important shift in the history of neoliberalism, where a 'justification for markets' has morphed into a 'justification for business', prioritizing the interests of (transnational) corporations (Davies, 2014, 38). In addition, as the above discussed *sensitive inversion* in EU policymaking suggests, next to being conceptualized as interest-led maximizers, economic agents are treated as sensitive beings too who need to be supported and protected.

In fact, the neoliberal 'justification for business' is no longer about the principles of economic competition that establish a level-playing field for business actors by limiting monopolistic tendencies. Rather, it is about pushing a 'competitiveness' agenda where the state is committed to supporting domestic businesses in order to augment their chances of success in the global market, vis-à-vis other, foreign business actors. Indeed, while in the EU's latest high-level digital strategy proclamation from 2020, *Shaping Europe's Digital Future,* references to competition appear in the context of establishing a level-playing field via regulatory tools (e.g. the Digital Markets Act), it is often with a view to boosting EU companies' productivity and global competitiveness by leveraging the benefits of the scale of the single market and *"providing them with an immensely powerful home market from which to launch themselves on the world stage"* (European Commission, 2020c, 8).

This resonates with the Arendt-inspired conceptual framework described above, which critiques the entrenched, modern imaginary of power that is about competitiveness and domination. Namely, whereas the competition discourse is conceptualized as establishing a level-playing field, thus aligning with horizontal imaginaries of power, the competitiveness discourse brings imaginaries of vertical power relations to the fore. This, when taken together with the desire to act alone and/or aspire to leadership while others follow, starts to fuel a discourse of power that is acted out on the global stage. As digital networks span the entire globe, it is of no surprise that the digital sovereignty discourse has emerged as a way *"to assert some form of collective control on digital content and/or infrastructures"* (Couture and Toupin, 2019, 4).

The prominence of the competitiveness agenda was confirmed by a critical discourse analysis that traced the implicit assumptions underlying the digital strategies of the European Union and the United States in the past 30 years, between 1984 and 2020 (Gulyás and Shahin, 2023). The research showed that both the EU and US digital strategies rest on a bedrock of neoliberal policymaking that prioritizes the fulfilment of business needs with the aim of boosting competitiveness and economic growth. The findings also confirmed the presence of the *sensitive inversion* in the field of digital policy. Catering to the needs of sensitive economic agents and treating technological development as the enabler of economic growth have laid the ground for a discourse on (global) competitiveness, which became a central element of digital policymaking in the EU and beyond. The depiction of the world as a competitive arena where agents act in order to win a race was found to be the baseline assumption for both European and American policymakers (Gulyás and Shahin, 2023). However, a shift away from this thinking has been indicated in the latest EU digital strategy, *Shaping Europe's digital future*, that emphasizes a more human-centered approach to digital policy. It does so by highlighting the importance of 'technology that works for the people' as well as by treating the green and digital transitions as twin priorities (European Commission, 2020c).

As the solutions to policy problems in the digital sphere often have global effects and repercussions (e.g. the EU's General Data Protection Regulation), examining the assumptions underlying them is important. After introducing the research methodology, in the next section I move on to examine the

European Commission's data strategy to understand the extent to which it is embedded in the above described modern conceptual framework. As data is critical to the development of the digital economy and society, and is expected to become a central element in the future of international trade, it is worth analysing the data strategy and policies of the EU both in domestic terms and vis-à-vis other players.

## Methodology

The analysis presented in the case study relies on the methodology of interpretive policy analysis. Like many other qualitative approaches, interpretive policy analysis builds on the assumption that language is not merely a transmitter but an active shaper of meanings. Accordingly, interpretive policy analysis aims to understand the meanings of policies by looking at how they are framed and what they represent for whom (Yanow, 2000). This can be done through tracing narratives, interpreting metaphors or critically analyzing discourses, which is the approach that this research takes.

In order to establish and analyze the conceptual link between digital sovereignty and the competitiveness agenda, the EU data policy field was chosen as a case study. Data is considered an essential resource for the future (global) economy and a key component of the overall digital transition in every respect. Hence, it is a central concern for EU digital sovereignty. The basis and conditions on which data will be conceived of, regulated, and traded in the future are currently being laid down in the EU and beyond. Therefore, it is important to examine the assumptions that recent data policies are built on and the meanings that these underlying assumptions carry and generate.

The research presented here traces the competitiveness discourse in policy documents that contain the underlying rationale for EU data policies. The first one is the overarching EU data strategy from 2020, titled *A European Strategy for data,* which describes the main concerns and direction of the policies addressing data. Policy strategy documents contain the overall vision and worldview of policymakers and therefore are ideal objects of analysis for uncovering otherwise hidden assumptions and meanings that the policies themselves rest on.

The other two policy documents examined are the impact assessment reports relating to the Data Governance Act (2020) and most recently to the Data Act (2022). Similarly, the types of justifications that give away the underlying rationales can be found in these kinds of policy documents that provide background knowledge, revealing the thinking on the basis of which policies are conceived. In this sense, they are again ideal resources for the kind of research put forth here, lending themselves much better to conceptually-grounded interpretive policy analysis than the texts of the regulations themselves would.

## EU data strategy and policies

In the European Union, several digital policy initiatives have been launched recently under the auspices of technological/digital sovereignty. One example is the regulation of data flows, where control over and access to data as a highly valuable asset and resource is regarded as a cornerstone of strengthening digital or technological sovereignty (European Commission, 2020a, 5).

The EU's overarching vision on the data economy was put forward in 2020 in its policy document titled *A European Strategy for data.* The strategy refers to data as the 'lifeblood' of economic development that is fundamental to European competitiveness, *"a potential source of growth and innovation that should be tapped"* (European Commission, 2020a, 1).

The strategy regards the suggested measures as contributing *"to a comprehensive approach to the data economy that aim to increase the use of, and demand for, data and data-enabled products and services throughout the Single Market"* (European Commission, 2020a, 1). By situating the measures in the context of the single market, and acknowledging that the purpose is to boost both the use of and demand for more data, the strategy caters to the logic of perpetuating endless growth. In fact, the strategy explains that *"currently, a small number of Big Tech firms hold a large part of the world's data"*, which *"could reduce the incentives for data-driven businesses to emerge, grow and innovate in the EU today."* Moreover, by proposing the creation of a single European data space, the strategy posits that *"the European data space will give businesses in the EU the possibility to build on the scale of the Single Market"*, indicating that the aim is for European companies to be more competitive on the global stage (*Ibid.*, 5).

These framings powerfully invoke the competitive race imaginary where the EU is in danger of lagging behind (*Ibid.*, 3). Indeed, the urgency and necessity are emphasized by stating that *"the sources of competitiveness for the next decades in the data economy are determined now. This is why the EU should act now"* (Ibid.). This framing corresponds to the Arendtian category of 'labor' discussed earlier: it is rooted in the notion of necessity as a mode of relation, indexing political action on survival and adaptation.

In the same section, the strategy's *"open but assertive approach to international data flows"* references the United States (US) and China as the EU's main competitors and positions itself in relation to them: on the one hand, as a wish to mitigate US tech companies' market concentration effects, on the other, as an alternative to China's lack of safeguards and protection of the data of individuals (Ibid.). The strategy emphasizes that between these two approaches, *"we have to find our European way"*, suggesting that there isn't only an economic race to be won, but also a regulatory race is taking place (*Ibid.*, 3).

The above findings demonstrate that the data strategy is to a great extent embedded in the modern imaginary resting on the centrality of vertical power relations, striving to win a race by dominating others. In the Arendtian framework, such conceptualizations of acquiring a leadership position correspond to 'work', i.e. the instrumentalizing mode of relation.

At the same time, the strategy also makes numerous references to the well-being of people, emphasizing the benefits of data for healthcare services, and also for advancing the green transition. These are however only mentioned in a domestic context as benefits for EU citizens. The global aspects of the data strategy are therefore mainly grounded in a desire to acquire a leadership position in the global digital economy, with the proliferation of data representing *"major opportunities for the EU to become a world leader in this area"* (European Commission, 2020a, 2).

It is important to note that in the EU data strategy there is an attempt to move away from a pure competitiveness agenda, towards a more human-centered discourse that begins to acknowledge people as sensitive agents. However, the strategy also asserts that data's *"potential should be put to work to address the needs of individuals and thus create value for the economy and society"*, demonstrating that even the fulfilment of people's needs is subordinated to the ultimate goal of economic value creation (European Commission, 2020a, 4). This again reaffirms the presence of the sensitive inversion: data is the 'lifeblood' of the personified economy, while people's needs are turned into functional assets to satisfy business purposes.

if we now move from the analysis of strategy to the analysis of policies, the two main regulations flowing from the data strategy, the Data Governance Act (passed in 2021) and the Data Act (political agreement reached in June 2023) testify to the EU's highly instrumental approach to power in the global digital policy space. The European Commission describes the two legislations on its website as the following: *"While the Data Governance Regulation creates the processes and structures to facilitate data, the Data Act clarifies who can create value from data and under which conditions"* (European Commission, 2022b).

Indeed, the Data Governance Act aims to boost data sharing among European stakeholders with the objective of propelling innovation through improving access to data. In the impact assessment report accompanying the policy proposal on data governance, it is apparent that the regulation is also driven by fears of non-EU companies dominating the European data market (European Commission, 2020b, 8). Thus, the Data Governance Act can be thought of as a tool to create a level-playing field for domestic companies in the name of fair competition. On the other hand, the impact assessment report for the Data Act makes it clear that the underlying driver of the proposal is first and foremost the fear of the *"loss of EU competitiveness in the global data economy"* (European Commission, 2022c, 9). Indeed, the impact assessment lays out a problem tree (figure 1) that names the problems and drivers that underpin the Data Act, as well as the consequences of inaction. Here, as shown in the graph below, the loss of EU competitiveness on the global stage is identified as the central consequence of not acting, revealing that this is the deep-seated anxiety that in fact drives the EU's data policymaking efforts.

**Figure 1: Data Act: Problem tree (European Commission, 2022c, 9)**

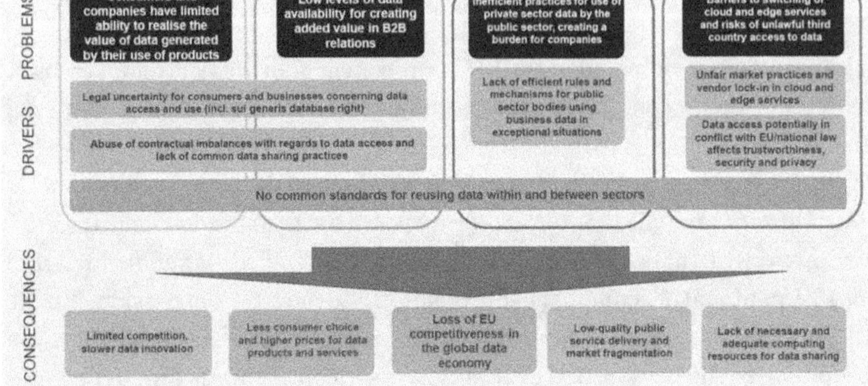

This finding reaffirms that on the global stage, the logic of competitiveness is at the center of EU digital policymaking, underlying its instrumental mode of relating to other actors merely as competitors that need to be dominated. What's more, it also makes it possible to frame the need for action as a necessity.

The discourse that has emerged in the EU's data policy can thus be understood as a symptom of the race mentality that neoliberal economic imaginaries bring to policymaking. This happens through the overvaluation of being in control and on top of others, conceptualizing relations on a vertical scale, and giving importance to acting alone, independently, as if that's where freedom was to be experienced. Going back to the Arendtian distinction of labor-work-action, and the resulting modes of relation (necessity-instrumentality-freedom), it can be argued that embedding policies in instrumentalizing conceptualizations of power would correspond to the domain of 'work', where one can remain a master by acting upon others as objects. What's more, when policies are framed in terms of being necessities, they are further reduced to the domain of 'labor', always limited to survival and adaptation.

## The competitiveness agenda: an obstacle to meaningful policies and co-operation

As N. Dewandre (2018) argues using the Arendtian conceptual framework, freedom is to be experienced in the domain of 'action', and not in the domains of 'labor' or 'work' that are grounded in necessity and instrumentality, respectively. It is in the realm of action where not rational subjects, but relational selves act together in plurality, recognizing each other as equals and playing out their distinct identities. For this, however, an important prerequisite is that relational selves need to have a sense of satiety: without it, they are permanently stuck in the same never-ending process of striving for more: competitiveness and control can give a meaning only in the absence of such a sense of satiety. In addition to satiety, the EU would also benefit from giving more room to experiencing vulnerability, which is in this sense the opposite of competitiveness. Acknowledging that the digital domain gives rise to new kinds of vulnerabilities that are not possible to control fully (e.g. disinformation, the dependence on others for raw materials, risks of malicious intentions or unintended breakdowns) could make room for policies that are more rooted in resilience, rather than control (Dewandre, 2015, 201-202).

The above case study is also a good illustration of William Davies' (2014) argument that the primary function of government has become the management of its competitive standing in comparison to other countries or regions. Davies explains how through the development of benchmarking, business strategy and management that were previously applied for the efficient running of firms, found their way to liberal economics and eventually to the neoliberal state starting from the 1970s. Benchmarking allowed for constant comparison with other states by creating a ranking and provided justification for policy action, regardless of whether states were in economic competition with each other or not. Important to this shift was also the figure of the entrepreneur as competition became indexed on the capacity to innovate (Davies, 2014).

The consequence of this was that soon firms and sovereign states started to be represented as *'ontologically equivalent and symbiotically related'* (Davies, 2014, 122).National economic prosperity depended on the performance of key industries and firms, whose success also depended on the conditions and regulations created by policymakers. As such, *"a new type of political-economic contest is constructed and represented, which lies between market liberalism (positive sum market competition between firms) and political mercantilism (zero sum geopolitical competition between states). In this contest, firms are quasi-political actors, seeking to position themselves in strategically advantageous positions in the global economy: states are quasi-corporate actors, seeking to do the same thing* (*Ibid.*)."

In this sense, the abovementioned *sensitive inversion* present in EU policymaking is a logical stance to take for policymakers. This is even more so in the field of digital policy: as pointed out by Marion Fourcade and Jeffrey Gordon (2020) in their article on statecraft in the digital age, contemporary 'dataist states' rely on data for their own functioning and policymaking. However, the state often does not own the infrastructure to collect this data and hence is dependent on the domestic private sector for it, resulting in its 'seeing like a market' (Ibid.).

The analysis shows that the EU's data policies and digital sovereignty discourse can be understood as a desire for control and/or not being controlled: two sides of the same coin that crowds out imaginaries for horizontal power relations and experiencing plurality and freedom, at the expense of a loss of meaning for politics. This meaning in the 21st century could be reinstated by acknowledging that life in the EU for most people reaches levels above satiety, and directing efforts at taking the twinning of the digital and green transitions seriously.

Indeed, it seems that European data policies have little regard for the environmental limits that constrain the process of digitalization: with their underlying assumption of the necessity to collect and gather ever more data in order to exploit it for improving the competitiveness of the European economy and boosting its growth, the policies are in contradiction with the mindset that would be needed to facilitate the green transition by leaving behind the growth agenda. By switching from the standpoint of rational subjects to relational selves, the EU could overcome this trap and instead start thinking about what kind of data should or should not be collected to begin with in its digital policymaking efforts.

Through the examination of the nexus between economic competitiveness and digital sovereignty as informed by a fundamentally vertical understanding of power relations, the analysis presented here also has implications for the future of global digital cooperation. This does not only concern the green transition, but also the future of the international economic order as data trade will likely be critical to the form it takes (James, 2020). Holding to the current imaginaries of power and politics, the EU risks the failure to engage with other countries and regions as partners, i.e. as plural equals who are unique.

## Conclusion

In this article I argued that the digital sovereignty discourse is a continuation and a direct consequence and manifestation of entrenched ways of thinking in EU policymaking that prioritize economic thinking over human needs. Two far-reaching consequences of this were highlighted: 1) the conflation of the state and the market, i.e. public authority and the private sector in digital policymaking; and 2) the fact that the current conceptualizations of the digital and green transitions are contradictory and their twinning gives rise to tensions.

Digital sovereignty discourses continue to position EU global competitiveness as the primary goal of the EU's data strategy, and the European economy and businesses as its main benefactors. While other scholarly contributions point to a recent shift from a neoliberal to a neo-mercantilist approach in EU digital policymaking (e.g. Farrand and Carrapico, 2022; Schmitz and Seidl, 2023), this article emphasized that on the level of policy rationales a neoliberal approach rooted in competitiveness remains the norm. The evolving geopolitical context, i.e. the Snowden revelations and the rise of China, may warrant the use of neo-mercantilist policy tools, but the underlying goals of EU digital policy-making have not changed and are made explicit when state actors prioritize the needs and desires of domestic businesses, rather than those of people. The sovereignty discourse brings to the fore and deepens a long-standing desire for control and domination upon others, and a wish to be 'on top' and able to act alone in the digital domain. Using an Arendtian approach to the analysis of policy documents, the paper has shown that EU digital politics conceives of power relations on a vertical scale rooted in notions of necessity and instrumentality. This comes at the expense of creating meaningful policies that would build on acting together with others for the benefit of people who need support and protection to cope with the digital transition. Furthermore, the EU's current approach does not leave room for experiencing satiety and vulnerability, and therefore freedom. This, however, would be essential for political action that aims to meaningfully co-operate with others in plurality, where actors are perceived as equals who are also unique. In lieu of this, sovereignty remains an illusion that is mistaken for freedom.

The green and digital transitions are incompatible, framed as they are at the moment: if competitiveness, growth and control remain the rationale for digital policymaking, it goes against the essence of the green transition. What would be needed is a shift of mindset on the part of the EU as well as other international policy actors that would facilitate thinking in terms of horizontal power relations, acknowledging limits to economic growth. However, the analysis of the EU's data strategy has shown that while traces of such a shift towards a more human-centric policymaking are detectable on the level of policy strategies, the policies that make the strategy tangible are still grounded in a modern approach to politics and power, manifested in the desire for control and acting alone in order to win a race.

The research also contributes to the understanding of how and in what ways the state and economic dimensions of digital sovereignty intertwine, or rather, how the state dimension already carries within the economic dimension when it ties together the discourses on competitiveness, leadership, and independence.

In the current geopolitical context, actors would need to recognize and activate imaginaries for plurality in their own societies, and also more broadly in their relations with other international actors. If plurality has any chance of becoming a useful mechanism for politics, however, the EU needs to craft a new message to be sent, rather than drawing on and activating old ones (competition, growth and jobs) that do not serve this purpose.

If the twinning of the digital and green transitions is a truly desirable ambition, the EU's digital policies have to reflect this shift and offer new meanings by rethinking the underlying assumptions of (digital) policymaking and the very concepts that are used to justify these. Sovereignty and control in the digital space are an illusion, fed by a discourse of competitiveness that does not sufficiently acknowledge the notion of satiety: the fact that the EU is able to offer abundance to its citizens as it is. It should be acknowledged that the main policy challenges lie in tackling the climate crisis triggered by unsustainable ways of production and consumption. As they stand, digital technologies and digitalization processes contribute to this problematic situation, however, they are also simultaneously seen as the means to the solution due to their scalability. Further research examining the underlying assumptions of green policymaking would be useful to better understand and harness the twinning of the green and digital transitions in order to make the most of them for the benefit of people and the planet.

## Bibliography

**Arendt, Hannah and Canovan, Margaret** (1998), *The Human Condition*, Second Edition, Chicago, University of Chicago Press.

**Bellanova, Rocco, Carrapico, Helena and Duez, Denis** (2022), "Digital/ sovereignty and European security integration: an introduction", *European Security*, vol. 31 (3), p. 337-355, <DOI:10.1080/09662839.2022.2101887>.

**Bernstein, Richard J.** (2011), "Hannah Arendt's Reflections on Violence and Power", *Iris. European Journal of Philosophy and Public Debate*, vol. 3 (5), p. 3-30.

**Breton, Thierry** (2021), "The Geopolitics of Technology", Brussels, European Commission, 27 July, <https://ec.europa.eu/commission/commissioners/2019-2024/breton/announcements/geopolitics-technology_en>.

**Castells, Manuel** (1996), *The rise of the network society*, Chichester, Blackwell.

**Couture, Stephane and Toupin, Sophie** (2019), "What does the notion of "sovereignty" mean when referring to the digital?", *New Media & Society*, vol. 21 (10), p. 2305-2322.

**Creemers, Rogier J.E.H.** (2020), "China's conception of cyber sovereignty: rhetoric and realization", *in* Dennis Broeders and Bibi van den Berg (eds), *Governing Cyberspace: Behavior, Power, and Diplomacy*, Lanham, Rowman & Littlefield, p. 107-142.

**Davies, William** (2014), *The Limits of Neoliberalism: Authority, Sovereignty and the Logic of Competition*, Thousand Oaks, Sage Publications.

**Dewandre, Nicole** (2015), "Rethinking the Human Condition in a Hyperconnected Era: Why Freedom is Not About Sovereignty But About Beginnings", *in* Luciano Floridi (ed.), *The Online Manifesto*, Springer, Cham, p. 195-215, <https://doi.org/10.1007/978-3-319-04093-6_20>.

**Dewandre, Nicole** (2018), "Political Agents as Relational Selves: Rethinking EU Politics and Policy-Making with Hannah Arendt", *Philosophy Today*, vol. 62 (2), p. 493-519.

**Dewandre, Nicole and Gulyás, Orsolya** (2018), "Sensitive economic personae and functional human beings: A critical metaphor analysis of EU policy documents between 1985 and 2014", *Journal of Language and Politics*, vol. 17 (6), p. 831–857, <https://doi.org/10.1075/jlp.17068.dew>.

**European Commission** (2020a), "A European Strategy for Data", COM(2020) 66 final, Brussels, 19 February, <https://eur-lex.europa.eu/legal-content/EN/TXT/PDF/?uri=CELEX:52020DC0066&from=EN>.

**European Commission** (2020b), *Impact Assessment Report – Data Governance Act*, SWD(2020) 295 final, Brussels, 25 November, <https://digital-strategy.ec.europa.eu/en/library/impact-assessment-report-and-support-study-accompanying-proposal-regulation-data-governance>.

**European Commission** (2020c), "Shaping Europe's Digital Future", COM(2020) 67 final, Brussels, 19 February, <https://eur-lex.europa.eu/legal-content/EN/TXT/PDF/?uri=CELEX:52020DC0067&from=en>.

**European Commission** (2022a), *A Europe fit for the digital age*, European Commission, <https://ec.europa.eu/info/strategy/priorities-2019-2024/europe-fit-digital-age_en>.

**European Commission** (2022b), *Data Act*, <https://digital-strategy.ec.europa.eu/en/policies/data-act>.

**European Commission** (2022c), *Impact Assessment Report – Data Act*, SWD(2022) 34 final, Brussels, 23 February, <https://ec.europa.eu/info/law/better-regulation/have-your-say/initiatives/13045-Data-Act-amended-rules-on-the-legal-protection-of-databases_en>.

**Farrand, Benjamin and Carrapico, Helena** (2022), "Digital sovereignty and taking back control: from regulatory capitalism to regulatory mercantilism in EU cybersecurity", *European Security*, vol. 31 (3), p. 435-453, <DOI: 10.1080/09662839.2022.2102896>.

**Floridi, Luciano** (2020), "The Fight for Digital Sovereignty: What It Is, and Why It Matters, Especially for the EU", *Philosophy & Technology*, vol. 33 (3), p. 369-378, <https://doi.org/10.1007/s13347-020-00423-6>.

**Fourcade, Marion and Gordon, Jeffrey** (2020), "Learning Like a State: Statecraft in the Digital Age", *Journal of Law and Political Economy*, vol. 1 (1), p. 78-108.

**Gulyás, Orsolya and Shahin, Jamal** (2023), "The Age of Digital Interdependence? A Critical Discourse Analysis of EU, US and ASEAN Digital Strategies",ETEE Working Paper Number 3.

**Harvey, David** (2005), *A Brief History of Neoliberalism*, Oxford, Oxford University Press.

**Heidebrecht, Sebastian** (2023), "From Market Liberalism to Public Intervention: Digital Sovereignty and Changing European Union Digital Single Market Governance", *Journal of Common Market Studies*, <https://doi.org/10.1111/jcms.13488>.

**Holborow, Marnie** (2015), *Language and Neoliberalism*, London, Routledge.

**James, Deborah** (2020), "Digital Trade Rules: A Disastrous New Constitution for the Global Economy, written by and for Big Tech", *Rosa-Luxemburg-Stiftung* [Blog post], 8 July, <https://www.rosalux.eu/en/article/1742.digital-trade-rules.html>.

**Mansell, Robin** (2014), "Here Comes the Revolution – the European Digital Agenda", *in* Karen Donders, Caroline Pauwels and Jan Loisen (eds), *The Palgrave Handbook of European Media Policy*, London, Palgrave Macmillan, p. 202-217, <https://doi.org/10.1057/9781137032195_12>.

**Mirowski, Philip and Plehwe, Dieter** (2009), *The Road from Mont Pèlerin: The Making of the Neoliberal Thought Collective*, Cambridge, Mass., and London, Harvard University Press.

**Mueller, Milton** (2017), *Will the Internet Fragment? Sovereignty, Globalization and Cyberspace*, Cambridge, Polity.

**Obendiek, Anke** (2021), "Take back control? Digital sovereignty and a vision for Europe", Jacques Delors-Centre, Hertie School of Governance, Berlin, 12 May, <https://www.delorscentre.eu/en/publications/detail/publication/take-back-control-digital-sovereignty-and-a-vision-for-europe>.

**Pasquale, Frank** (2018), "From territorial to functional sovereignty: the case of Amazon", *OpenDemocracy*, 5 January, <https://www.opendemocracy.net/en/digitaliberties/from-territorial-to-functional-sovereignty-case-of-amazon/>.

**Pohle, Julia and Thiel, Thorsten** (2020), "Digital sovereignty", *Internet Policy Review*, vol. 9 (4), p. 1-19, <https://policyreview.info/concepts/digital-sovereignty>.

**Roberts, Huw, Cowls, Josh, Casolari, Federico, Morley, Jessica, Taddeo, Mariarosaria and Floridi, Luciano** (2021), "Safeguarding European values with digital sovereignty: an analysis of statements and policies", *Internet Policy Review*, vol. 10 (3), p. 1-26, <https://doi.org/10.14763/2021.3.1575>.

**Schmitz, Luuk and Seidl, Timo** (2023), "As Open as Possible, as Autonomous as Necessary: Understanding the Rise of Open Strategic Autonomy in EU Trade Policy", *Journal of Common Market Studies*, vol. 61 (3), p. 834-852, <https://doi.org/10.1111/jcms.13428>.

**Shahin, Jamal** (2007), "The reassertion of the state: governance and the information revolution", *in* Myriam Dunn, Sai Felicia Krishna-Hensel and Victor Mauer (eds), *The Resurgence of the State: Trends and Processes in Cyberspace Governance*, London, Ashgate, p. 9-34.

**Thumfart, Johannes** (2022), "The Norm Development of Digital Sovereignty between China, Russia, the EU and the US: From the Late 1990s to the Covid-Crisis 2020/21 as Catalytic Event", *in* Dara Hallinan, Ronald Leenes and Paul De Hert (eds), *Enforcing Rights in a Changing World. Computers Privacy Data Protection (CPDP) 14*, London, Hart Publishing, p. 1-44.

**Van Dijck, José, Poell, Thomas and De Waal, Martijn** (2018), *The platform society*, Oxford, Oxford University Press.

**Villa, Dana** (2000), "Introduction: The development of Arendt's political thought", *in* Dana Villa (ed.), *The Cambridge Companion to Hannah Arendt*, Cambridge, Cambridge University Press, p. 1-22.

**Yanow, Dvora** (2000), *Conducting interpretive policy analysis*, Thousand Oaks, Sage Publications, <https://doi.org/10.4135/9781412983747>.

**Orsolya Gulyás** is a PhD Researcher at the Brussels School of Governance, focusing on discourses in digital policymaking. Combining her backgrounds in public policy and media studies, she researches issues at the intersection of technology and society with an interest in critical theories and methodologies.

orsolya.gulyas@vub.be

POLITIQUE EUROPÉENNE
N° 81 | 2023
Julien Rossi
p. 54-85

**La structure argumentative d'un demi-siècle de politique européenne de protection des données à caractère personnel**

Malgré les profondes transformations intervenues depuis l'adoption des premières lois de protection des données personnelles dans les années 1970, tant sur le plan technique que sur le plan politique, les dispositions en vigueur dans les textes actuels, au premier desquels le Règlement général de protection des données adopté en 2016, en reprennent les principaux principes et définitions. Le présent article s'appuie sur les approches cognitives des politiques publiques pour montrer comment cette forte continuité s'explique en large part par la stabilité de la structure argumentative des débats au sein des arènes où s'élaborent ces textes.

**The Argumentative Structure of a Half-Century of European Data Protection Public Policy**

Profound technological and political changes have occurred since the adoption of the first data protection laws in the 1970's. Despite this, their general principles and definitions have remained largely unchanged in currently applicable texts, such as the General Data Protection Regulation, approved in 2016. This article adopts a cognitive approach to public policy to show that this strong continuity can be attributed in a large part to the stability of the argumentative structure of debates held within the institutional spaces where these legal norms are produced.

# La structure argumentative d'un demi-siècle de politique européenne de protection des données à caractère personnel

**Julien Rossi**
Université Paris 8, chercheur au CÉMTI, et chercheur associé au COSTECH, Université de technologie de Compiègne

L'adoption du Règlement général de protection des données[1] (RGPD), entérinée en 2016, a réformé le droit de l'Union en matière de protection des données à caractère personnel, qui reposait jusqu'alors sur une directive remontant à 1995[2]. L'une des explications avancées pour comprendre cette réforme est le changement technique intervenu, et notamment la diffusion généralisée des usages d'Internet, du web social, et du traitement massif de données personnelles (Goyal, Howlett et Taeihagh, 2021). La mise à l'agenda de cette réforme peut être partiellement expliquée par ces changements techniques, mais le RGPD marque-t-il une réelle rupture avec l'état antérieur du droit ? Au-delà de la rhétorique de la « compliance » (Favro, 2017) et de l'approche « basée sur le risque » (Quelle, 2017), il semblerait que le contenu du droit des données personnelles, loin d'avoir été « disrupté » par la technologie, soit resté essentiellement stable (Bennett et Raab, 2020 ; Gellert, 2016). Comment expliquer cette stabilité ?

La toute première loi dite de « protection des données », de l'allemand « *Datenschutz* », a été adoptée par le Land de Hesse en 1970 : l'année qui a suivi la mise en service du réseau Arpanet, ancêtre d'Internet, plus de dix ans avant la commercialisation de l'IBM PC, et plus de vingt ans avant l'ouverture d'Internet à un usage commercial. C'est dire si le milieu technique dans lequel se situent les débats politiques sur le droit à la vie privée a en effet changé. Or, de nombreuses études sur la vie privée ont montré l'importance que jouaient

1 Règlement (UE) 2016/679 du Parlement européen et du Conseil du 27 avril 2016 relatif à la protection des personnes physiques à l'égard du traitement des données à caractère personnel et à la libre circulation de ces données, et abrogeant la directive 95/46/CE (règlement général sur la protection des données).

2 Directive 95/46/CE du Parlement européen et du Conseil, du 24 octobre 1995, relative à la protection des personnes physiques à l'égard du traitement des données à caractère personnel et à la libre circulation de ces données.

des éléments d'infrastructure, comme l'invention de la chambre à coucher ou l'agrandissement des villes, sur l'évolution du rapport entre espaces privés et publics (Guerrand, 1987 ; Mumford, 1938 ; Prost, 1987). L'invention de la photographie instantanée avait, de son temps, inspiré aux juristes états-uniens Samuel Warren et Louis Brandeis (1890) l'idée qu'il serait nécessaire de se protéger de son intrusion en reconnaissant l'existence d'un droit nouveau : le droit à la vie privée, défini comme le droit d'être laissé tranquille (*right to be let alone*) (Glancy, 1979). Dans les années 1960 et 1970, l'apparition de l'ordinateur électronique avait été un des principaux facteurs ayant conduit à un retour en force de la question du droit à la vie privée à l'agenda politique d'abord des États-Unis, puis des pays d'Europe de l'Ouest (Holvast, 2013 ; Miller, 1971 ; Packard, 1965 ; Westin, 1967).

La période a aussi été marquée par de profondes transformations politiques. En 1970, ni les crises pétrolières, ni le tournant néolibéral, ni la chute de l'URSS, les attentats du 11 septembre ou encore la crise des subprimes n'avaient encore eu lieu. L'Union européenne n'était qu'un ensemble de Communautés européennes, dont les commissions venaient tout juste de fusionner.

Toutes ces raisons auraient en effet justifié que l'on s'attende à ce que le contenu du RGPD, adopté le 27 avril 2016 par le Parlement européen et le Conseil, marque une rupture par rapport à la directive de 1995 qu'il abroge, et *a fortiori* par rapport aux premières lois nationales adoptées en Europe à partir du début des années 1970. Les crises et transformations majeures externes à un secteur de l'action publique sont en effet réputées être une condition nécessaire, quoique non suffisante, pour conduire à un changement (Sabatier, 1998). De nombreux discours produits pendant la période de négociation du RGPD, entre 2009, date de l'annonce d'un projet de réforme par la Commission, et 2016, année de l'adoption du texte, ont justifié la nécessité de celui-ci par ces transformations socio-techniques :

> « Le monde a profondément changé depuis 1995, l'année d'adoption du cadre en vigueur en matière de protection des données de l'UE. Nous vivons à présent dans un monde offrant une multitude de possibilités de communication. [...] Cette évolution a des implications considérables pour nos économies. [...] Mais cela soulève également un certain nombre de questions non négligeables concernant les droits des citoyens. Comment pouvons-nous assurer la protection des données dans un monde de connectivité tous azimuts ? Comment pouvons-nous

renforcer la confiance des consommateurs dans un monde confronté à une véritable explosion des volumes de données ? » (Discours de Viviane Reding, prononcé le 7 mars 2013[3]).

L'adoption d'un règlement en remplacement d'une directive, s'inscrit dans un processus de renforcement du transfert vers l'Union de compétences en matière de protection des données (Karaboga, 2018 ; Lynskey, 2016). Une hypothèse qui semble alors raisonnable serait de supposer que l'européanisation de la protection des données s'est opérée sous l'impulsion de la Commission européenne, par la méthode communautaire, à partir de la proposition de la directive sur la protection des données présentée en 1990[4] et adoptée en 1995. Or, l'existence d'une Convention du Conseil de l'Europe datant de 1981, et le fait, constaté dans les travaux de Colin Bennett (1992) ainsi que par Priscilla Regan (1995), que les différentes lois adoptées en Europe au cours des années 1970 comportaient dès l'origine de nombreux points communs, sont le signe d'une coordination européenne précoce en matière de protection des données. Celle-ci a précédé de plus d'une décennie l'implication des institutions communautaires. Ainsi, l'émergence d'un droit fondamental nouveau, le droit à la protection des données à caractère personnel (Hondius, 1975), et la réflexion autour des mesures concrètes lui donnant corps, se sont organisées autour de deux groupes d'experts, l'un majoritairement, l'autre entièrement européen, à l'OCDE et au Conseil de l'Europe. Ce n'est que par la suite, sous l'impulsion des premières autorités nationales de protection des données, regroupées en un réseau transgouvernemental structuré par des rencontres régulières à l'occasion de l'International Conference of Data Protection and Privacy Commissioners (ICDPPC) dont la première édition a lieu à Bonn en 1979, que la politique européenne de protection des données est devenue communautaire (Eberlein et Newman, 2006 ; Newman, 2008). L'abondante littérature que nous venons de citer montre que contrairement à ce que semble suggérer la lecture de certains travaux plus récents[5], le RGPD est une étape supplémentaire dans un processus d'européanisation déjà ancien.

3 Discours de V. Reding prononcé le 7 mars 2013 à la 2e édition de la Conférence européenne annuelle sur l'informatique en nuage à Bruxelles. Disponible en ligne : <https://ec.europa.eu/commission/presscorner/detail/fr/SPEECH_13_197>.

4 Proposition de directive du Conseil relative à la protection des personnes à l'égard du traitement des données à caractère personnel, présentée par la Commission le 27 juillet 1990 - COM (90) 314 FINAL - SYN 287.

5 C'est le cas par exemple de l'article dans lequel Orla Lynskey (2016) expose par ailleurs une analyse poussée et argumentée des tensions générées par la transformation du Groupe de travail de l'article 29 en Comité européen de protection des données, notamment au regard du principe de subsidiarité.

De ces observations liminaires découlent une question et une contrainte méthodologique qui structurent cet article. La question est la suivante : compte tenu de ce que nous avons exposé, si le contenu du droit de l'Union sur les données personnelles est resté stable, comment expliquer cette stabilité ? Quant à la contrainte, il s'agit de devoir tenir compte non seulement de l'action de l'UE et de la Communauté économique européenne (CEE) qui l'a précédée, mais également des autres arènes européennes ou majoritairement européennes au sein desquelles s'est structuré à l'échelle européenne ce champ de l'action publique, dès le début des années 1970. Cette contrainte peut être surmontée en adoptant un point de vue qui ne soit pas centré sur une institution en particulier, mais sur des coalitions d'acteurs et actrices agissant au sein d'un champ de l'action publique (Dubois, 2014), ou un sous-système de politique publique (Sabatier, 1993), dont le périmètre peut inclure des acteurs et actrices évoluant au sein d'une grande diversité d'organisations.

Le droit des données personnelles contient un grand nombre de dispositions, que nous n'avons pas la place d'examiner en détail dans l'espace d'un article. Nous nous focaliserons sur un enjeu crucial : les débats au sujet de la définition de la notion de « donnée à caractère personnel ». Celle-ci concentre en effet de nombreux enjeux, le premier d'entre eux étant de déterminer le champ d'application matériel de toutes les autres dispositions contenues dans le droit des données personnelles. Cette définition s'est stabilisée au cours des négociations sur la Convention 108 du Conseil de l'Europe, adoptée en 1981, et a fait l'objet de tentatives de remise en cause fondamentale durant les négociations sur le RGPD entre 2009 – date des premières consultations réalisées par la Commission – et 2016.

Nous écarterons la piste de la dépendance au sentier comme principal facteur d'explication. En effet, la saillance de la question de la protection des données, renforcée par les révélations Snowden (Laurer et Seidl, 2021 ; Musiani, 2015 ; Rossi, 2018), a conduit à un nombre record d'amendements déposés au Parlement européen : pas moins de 3999 amendements (Laurer et Seidl, 2021). Certains – nous le verrons en ce qui concerne la définition de la notion de donnée personnelle, et donc du champ d'application matériel – proposaient des remises en cause en profondeur du contenu de ce domaine du droit. De ce fait, l'adoption du RGPD semblerait de prime abord illustrer le phénomène de dépendance au sentier.

Nous rejoignons toutefois la critique adressée par Adrian Kay à ce concept, selon laquelle il ne permet pas toujours de « comprendre ou expliquer ce qu'il désigne ; les processus de dépendance au sentier, même une fois identifiés, nécessitent une analyse théorique[6] » (Kay, 2005, 554). En effet, pourquoi l'héritage des décisions passées a-t-il été plus fort que la conjonction entre les changements politiques, technologiques, et le fort investissement de groupes d'intérêts favorables à un changement de fond ? Cette question se pose particulièrement dans le contexte des débats sur le RGPD, puisque malgré l'absence de remise en cause des fondamentaux de la politique européenne de protection des données, l'augmentation massive du montant des sanctions (à l'article 83), l'introduction de nouveaux mécanismes de coopération entre autorités de contrôle (à l'article 60) ou encore l'importation dans le droit de l'Union (Bennett et Raab, 2020) de l'évaluation d'impact relative à la protection des données (aux articles 35 et 36) ont tout de même généré des coûts d'investissement et d'apprentissage prévisibles pour les acteurs concernés. Si l'on suit la théorie de la dépendance au sentier, celle-ci explique la résistance au changement en large partie par des considérations économiques de coûts d'adaptation croissants (Pierson, 2000). Or, en matière de protection des données, un changement radical tel que la réduction du champ d'application matériel aux seules données directement identifiables, tel qu'il a été proposé par une coalition d'acteurs lors des négociations sur le RGPD, aurait engendré des économies au lieu de générer de nouveaux coûts d'adaptation à un ensemble de règles qui étaient certes en vigueur depuis longtemps, mais très inégalement appliquées du fait de la faiblesse du montant des sanctions.

Il nous semble donc nécessaire d'aller chercher ailleurs que dans la théorie de la dépendance au sentier l'explication de cette stabilité dans les paramètres fondamentaux de la politique de l'Union européenne en matière de protection des données. Ainsi, après avoir présenté la façon dont s'est construite la politique européenne de protection des données, nous étudierons donc la structure argumentative au cours de cette première phase, entre 1970 et 1981, puis au cours du processus d'adoption du RGPD, entre 2009 et 2016. Nous analyserons les matériaux empiriques présentés à l'aune des approches cognitives des politiques publiques, dans une discussion théorique qui permettra notamment de revenir sur la notion de structure argumentative des politiques publiques que nous proposons. Nous nous appuierons sur du matériau collecté dans le cadre d'un travail de thèse effectué entre 2015 et 2020 (Rossi, 2020), au cours duquel nous avions notamment réalisé un travail d'analyse qualitative de documents (Bowen, 2009) sur un corpus de 534 *position papers* rédigés dans

6 Notre traduction depuis l'anglais. Texte original : « provide necessary or sufficient conditions to understand or explain that which it labels: path dependent processes, even when identified, require theorizing. »

le cadre des débats sur le RGPD, de documents d'archives issus principalement de l'OCDE et du Conseil de l'Europe, et d'une vingtaine d'entretiens réalisés avec des acteurs des politiques européennes de protection des données, certains depuis le début des années 1970.

## La construction d'une politique européenne de protection des données à caractère personnel sans, puis avec les institutions communautaires

La toute première loi dite de « protection des données », de l'allemand *Datenschutz*, a été adoptée par le Landtag du Land de Hesse en 1970[7]. La loi sur les données de Suède est adoptée dès 1973[8], et la France suit avec sa loi informatique et libertés en 1978[9]. Or, ces différentes lois comprenaient de nombreux points communs, parmi lesquels l'instauration d'une autorité administrative indépendante chargée de veiller au respect de ces nouvelles règles, l'instauration d'un droit d'accès des personnes physiques aux données qui les concernent, et, pour les deux dernières, une définition des notions de « donnée personnelle » (« *personuppgift* ») et de « donnée nominative » identiques sur le fond.

Bien que ce soit en Europe que les premières lois de protection des données soient apparues, les premières réflexions politiques et juridiques sur les effets de l'informatisation sur le droit à la vie privée viennent des États-Unis d'Amérique, qui sortent, à la fin des années 1950 et au début des années 1960, de plusieurs années de surveillance intense soutenue notamment par l'action du sénateur Joseph McCarthy, président du Comité sur les Opérations gouvernementales du Sénat des États-Unis. La réaction de journalistes (Packard, 1965), de juristes (Westin, 1967), de député·e·s (Chambre des représentants des États-Unis d'Amérique, 1966) et de sénateurs (Sénat des États-Unis d'Amérique, 1967), combinée aux prises de position de statisticiens et d'informaticiens des années 1950 s'inquiétant du risque que l'automatisation du traitement d'informations personnelles faisait peser sur la vie privée (Ware, 1967), a conduit à de nombreuses publications qui ont été diffusées en Europe. Ces réflexions étaient notamment alimentées par une vision libérale de la vie privée, qui devait servir à l'épanouissement de l'individu (Westin, 1967), et aussi une image de l'ordinateur comme machine menaçante liée au complexe

7 Hessische Datenschutzgesetz, 7 octobre 1970.

8 Datalag (1973:289).

9 Loi n° 76-18 du 6 janvier 1978 relative à l'informatique, aux fichiers et aux libertés.

militaro-industriel. L'imaginaire techno-utopiste décrit dans les travaux de Fred Turner (2008) et de Patrice Flichy (2001) n'existait pas encore, ou était extrêmement marginal.

Spiros Simitis, l'un des auteurs de la première loi de protection des données, adoptée en Hesse en 1970, nous indiquait ainsi en entretien que « l'une des principales sources d'inspiration pour notre propre réflexion et prospective était une étude continue de la jurisprudence aux États-Unis concernant l'automatisation. Les Américains [...] ont fourni la majeure partie des bases de réflexion[10]. » Un exemple de la façon dont les débats outre-Atlantique ont été importés en Europe est la trajectoire de Peter Hustinx, qui fut de 2004 à 2014 Contrôleur européen de la protection des données, l'autorité qui veille au respect des règles de protection des données par les institutions de l'Union. Celui-ci avait été, jeune fonctionnaire du Ministère de la Justice aux Pays-Bas, désigné pour participer aux travaux d'une commission sur le sujet au retour d'un voyage d'études qu'il avait effectué aux États-Unis, et où il avait découvert les réflexions sur le droit de l'informatique[11].

Le plus ancien rapport officiel français sur le sujet dont nous ayons connaissance, une étude du Conseil d'État de 1970, évoque aussi plusieurs de ces travaux provenant d'outre-Atlantique. Ce rapport mentionne aussi des travaux et réflexions en cours dans d'autres pays européens, montrant l'existence d'une communication sur ces sujets entre haut·e·s fonctionnaires des pays du bloc de l'Ouest (Conseil d'État, 1970, 6). Il précède de plusieurs années l'article de Philippe Boucher (Boucher, 1974) dans *Le Monde* sur l'affaire « SAFARI », projet de fichier de données personnelles du ministère de l'Intérieur, souvent présentée comme l'origine de la réflexion sur le rapport entre informatique et libertés en France (Vitalis, 2010).

Le groupe d'experts sur la protection de la vie privée vis-à-vis des banques de données électronique du Conseil de l'Europe, dont la première réunion se tient en 1972[12], et le groupe d'experts sur l'utilisation des ordinateurs de l'OCDE, dont les premières publications remontent à 1971 (voir : Niblett, 1971), ont été deux arènes qui ont étroitement coopéré et dans lesquelles se sont rencontrées

10 Entretien avec S. Simitis du 27 septembre 2016, cité dans Rossi, 2020, p. 188-189. Traduit de l'anglais. Texte original : « one of the main sources of inspiration of our reflection and of our expectations, was a continuous study on decisions, court decisions, in the United States, that had to do with automation. The Americans [...] offered most of the material, »

11 Entretien avec P. Hustinx du 15 juin 2016.

12 Conseil de l'Europe, EXP/Prot.Priv./EDB (72) 5.

les personnes qui allaient bientôt se constituer en réseau transgouvernemental de haut·e·s fonctionnaires soucieux de faire adopter dans leurs pays des règles sur la protection des données à caractère personnel, au sujet desquelles ils avaient formé un consensus au niveau transnational (Bennett, 1992, 127). Ont participé à ces groupes des personnes comme P. Hustinx, que nous avons déjà évoqué, Louis Joinet, connu pour avoir contribué à la création de la Commission nationale de l'informatique et des libertés (CNIL), ou encore Stefano Rodotà, qui fut entre 1997 et 2005 le premier président de l'Autorité italienne de protection de la vie privée.

La mise à l'agenda par le haut de la problématique de la protection des données s'explique notamment par le fait que, comme le rappelle P. Regan, « au moment où les questions de protection de la vie privée ont été mises à l'agenda public dans les années 1960, aucune communauté dédiée à la protection de la vie privée [*privacy community*] impliquée dans le travail législatif ne s'était encore formée. » Or, dès « avant la fin des années 1970, un noyau de communauté politique [policy community] s'intéressant de façon générale à la vie privée s'était mis en place [...][13] » (Regan, 1995, 20-21).

Ainsi, comme le constatait dès 1980 Michael Kirby, le président du groupe d'experts de l'OCDE chargé d'établir les Lignes directrices régissant la protection de la vie privée et les flux transfrontières de données de caractère personnel du 23 septembre 1980 :

> « Au vu des différents langages, traditions juridiques, et valeurs culturelles et sociales, des désaccords fondamentaux auraient été attendus. Cependant, curieusement, dans tous les travaux internationaux majeurs effectués jusqu'ici sur la question, un large consensus a été trouvé quant aux « règles de base » fondant la législation de protection de la vie privée dans les droits nationaux[14] » (Kirby, 1980, 29, cité par Bennett, 1992, 95).

Concrètement, ces convictions se sont traduites par l'adoption de textes qui ne s'appliquaient pas seulement à un ensemble d'informations considérées

13 Traduit de l'anglais. Texte original : « By the late 1970's [...] a core policy community interested in general privacy existed [...]. »

14 Traduction. Texte original : « Given the different langages, legal traditions, and cultural and social values, fundamental disagreements might have been expected. Surprisingly, in all of the major international efforts that have so far addressed this problem, there has been a broad measure of agreement on the "basic rules" around which domestic privacy legislation should cluster. »

par avance par le législateur comme étant de nature privée, mais à l'ensemble des données relatives à un individu. Ainsi, l'article 2 (a) de la Convention 108 du Conseil de l'Europe[15], adoptée en 1981, définit la notion de donnée à caractère personnelle comme « toute donnée concernant une personne physique identifiée ou identifiable ». En présence d'un traitement automatisé en tout ou partie, ou de fichiers structurés de telles données, ce traité, toujours en vigueur aujourd'hui, prévoit l'application d'un certain nombre de principes. Ces derniers, énumérés à l'article 5, sont la loyauté et la licéité, la limitation des finalités, la minimisation, la qualité et l'exactitude, et la limite de la durée de conservation. L'article 7 pose également un principe de sécurité et de confidentialité. Ces principes sont, pour l'essentiel, des reformulations d'idées proposées à l'origine dans un rapport de la British Computer Society de 1971. Ils furent repris par un rapport remis en 1972 au gouvernement britannique (Younger, 1972), dont un certain G. P. Pratt, membre de la commission l'ayant rédigé, a présenté les conclusions au groupe d'experts du Conseil de l'Europe, favorisant la diffusion de ces idées à l'ensemble de l'Europe.

Cette convergence précoce à l'échelle européenne s'est ainsi faite en dehors des institutions communautaires, et même en dépit du refus de la Commission européenne, exprimé à maintes reprises dans les années 1970 et 1980, d'intervenir pour harmoniser le droit naissant des données à caractère personnel. Cette résistance n'a été vaincue que par la détermination des autorités de protection des données instituées dans quelques pays européens, déjà regroupées au sein de l'ICDPPC. C'est dans ce cadre que les présidents de ces autorités prirent la décision de menacer la Commission d'entraver la libre circulation des données personnelles au sein du marché intérieur, au risque de faire échouer la mise en place du Système d'information Schengen (SIS), jugé indispensable au bon fonctionnement du projet d'Espace Schengen (Newman, 2008).

En 1989, la CNIL française fit usage de son pouvoir réglementaire pour interdire le transfert de données sur les employés de la société Fiat depuis la France vers le siège de l'entreprise en Italie[16]. En 1990, S. Simitis, alors commissaire à la protection des données du Land de Hesse, menaçait ainsi ouvertement, à défaut d'une harmonisation des règles de protection des données, d'utiliser les compétences de l'autorité qu'il dirigeait pour empêcher la réalisation du système d'information Schengen (Newman, 2008, p. 89).

15 Convention pour la protection des personnes à l'égard du traitement automatisé des données à caractère personnel.

16 CNIL, délibération 89-78 du 11 juillet 1989.

La même année, la Commission européenne publiait deux propositions de directives sur la protection des données à caractère personnel[17] et sur la protection de la vie privée dans les communications électroniques[18], adoptées respectivement en 1995 et 1997. De façon cohérente avec le paradigme libéral de la vie privée partagé par la communauté des défenseurs de la vie privée, la directive de 1995 consacrait par son article 7 l'importance accordée au consentement de l'individu, en haut de la liste des nouveaux critères de licéité, bases légales dont tout traitement de données à caractère personnel devait désormais disposer pour être considéré comme licite. Ainsi, ce réseau d'entrepreneurs·ses transgouvernementaux·les (Eberlein et Newman, 2006) a obtenu la communautarisation d'une politique publique qui était toutefois déjà structurée à l'échelle européenne. Comme le souligne Abraham Newman (2008), le récit de la communautarisation de la protection des données en Europe ne correspond ni au récit intergouvernementaliste (Moravcsik, 1991) reposant sur une convergence entre les intérêts des États-membres, ni au récit néofonctionnaliste (Haas, 1958) qui suppose une demande issue d'acteurs privés, mais sur ce qu'il appelle un phénomène d'entrepreneuriat transgouvernemental.

Dorénavant, la place de l'UE dans le pilotage d'une action publique européenne harmonisée de protection des données à caractère personnel et de libre circulation de celles-ci s'illustre par la reconnaissance institutionnelle du réseau transgouvernemental constitué des autorités nationales de protection des données, d'abord par l'article 29 de la directive de 1995, puis aujourd'hui par l'article 68, non plus d'une directive, mais du Règlement général de protection des données, par l'institution d'un Contrôleur européen de la protection des données, par celle d'un groupe d'expert·e·s dédié au sein du Conseil de l'UE (le DAPIX), par la reconnaissance d'un droit à la protection des données distinct du droit à la vie privée à l'article 8 de la Charte des droits fondamentaux de l'Union européenne, et enfin par l'existence d'un article 16 TFUE qui, depuis le traité de Lisbonne, a entièrement transféré à l'Union, chose exceptionnelle en matière de droits humains, les compétences en matière de protection des données. Il existe même une directive qui harmonise les règles de traitement des données traitées par les autorités compétentes à des fins de prévention

17 Proposition de directive du Conseil relative à la protection des personnes à l'égard du traitement des données à caractère personnel. COM (90) 314 FINAL - SYN 287.

18 Proposition de directive du Conseil concernant la protection des données à caractère personnel et de la vie privée dans le cadre des réseaux numériques publics de télécommunications, en particulier des réseaux numériques à intégration de services (RNIS) et des réseaux mobiles numériques publics. COM (90) 314 FINAL - SYN 288.

et de détection des infractions pénales, d'enquêtes et de poursuites en la matière ou d'exécution de sanctions pénales[19]. Ces institutions et ces textes coexistent avec la Convention 108 du Conseil de l'Europe, qui remonte à 1981 et a été actualisée en 2018, et son comité consultatif, élargi depuis à des pays non européens, signataires de la convention.

## Entre 1970 et 1981, convaincre de la nécessité de créer une nouvelle catégorie de droit : le droit des données à caractère personnel

Les personnes qui ont contribué à la rédaction des premiers rapports puis des premiers textes consacrant un droit à la protection des données à caractère personnel partageaient une compréhension commune d'un droit à la vie privée perçu comme un droit de l'individu au contrôle de l'accès à soi et aux informations qui le concernent. Cette conception, qualifiée de « paradigme libéral de la vie privée » par Charles Raab et C. Bennett (Bennett et Raab, 2003), est notamment l'œuvre de théoricien·ne·s du droit tels qu'Alan Westin, qui a participé à certains travaux de l'OCDE et qui est régulièrement cité dans des rapports officiels de l'époque (voir par exemple : Gallouédec Genuys et Maisl, 1976 ; Koopmans, 1974 ; Niblett, 1971 ; OCDE, 1974 ; Chambre des représentants des États-Unis, 1966).

Les différents documents d'archives consultés donnent accès à des prototypes de définitions de la notion de « donnée à caractère personnel » qui laissent entrevoir deux hésitations : concerne-t-elle uniquement les personnes physiques, ou bien également les personnes morales ? Faut-il limiter cette définition aux seules données relevant de la sphère privée ou intime, ou bien faut-il appliquer les règles en cours de réflexion à toutes les données relatives

19 Directive 2016/680/UE du Parlement européen et du Conseil du 27 avril 2016 relative à la protection des personnes physiques à l'égard du traitement des données à caractère personnel par les autorités compétentes à des fins de prévention et de détection des infractions pénales, d'enquêtes et de poursuites en la matière ou d'exécution de sanctions pénales, et à la libre circulation de ces données, et abrogeant la décision-cadre 2008/977/JAI du Conseil.

à une personne physique identifiée ou identifiable ? Les extraits sélectionnés ci-dessous témoignent de ces hésitations initiales :

> « Énoncé de la forme – nom individuel, énoncé factuel. L'énoncé factuel peut être implicite ou donné par le contexte. Par exemple, si une bande magnétique contient uniquement une liste de noms, et qu'il est indiqué qu'elle contient seulement une liste de prisonniers. L'identité individuelle peut être submergée dans des statistiques apparemment anodines, mais si elle peut être extraite, alors il faut considérer qu'elle est présente[20] » (British Computer Society, 1972, 12).
> « Données concernant une personne physique ou morale[21] » (Projet de résolution relative à la protection de la vie privée vis-à-vis des banques de données électroniques dans le secteur privé, document EXP/Prot.Priv./EDB (72) 14 du Conseil de l'Europe, 1972).
> « Données [...] relatives à la vie privée ou à l'intimité des individus qu'elles concernent, en particulier les informations concernant leur race, religion, opinions politiques, mœurs, santé ou casier judiciaire[22] » (Proposition de Jean-Paul Costa dans l'annexe V du document EXP/Prot.Priv./EDB (72) 17 du Conseil de l'Europe, 1972).

En définitive, l'option qui s'est assez vite imposée sur les autres est que la protection des données relevait de la catégorie des droits humains et des droits de la personnalité. Cela s'est traduit dans la Convention 108 par la définition suivante : est une donnée personnelle « toute information concernant une personne physique identifiée ou identifiable » (art. 2 (a) de la Convention 108).

De plus, les personnes qui ont participé aux groupes d'expert·e·s de l'OCDE et du Conseil de l'Europe ont vite compris que l'une des raisons pour lesquelles l'ordinateur représentait un risque à l'égard de la vie privée, c'était parce que la combinaison, grandement facilitée par la technique, de données d'apparence anodines pouvait faire émerger des informations de nature plus sensible. Ainsi,

20 Traduit de l'anglais. Texte original : « Statement of the form - Individual name, factual statement. The factual statement may be implied or given by context: for example, if a magnetic tape containing only a list of names is said to be a list of prison inmates. The individual identity may be submerged in apparently anonymous statistics, but if it can be extracted then it must be considered to be there. »

21 Traduit de l'anglais. Texte original : « data concerning physical or legal persons »

22 Traduit de l'anglais. Texte original : « data [...] as relates to the private life or privacy of individuals whom it concerns, and particularly information concerning their race, religion, political opinions, morals, health or past judicial record »

selon le juriste italien S. Rodotà, qui participait à une conférence de l'OCDE sur le sujet en 1974 :

> « Il devient de plus en plus difficile de définir le type d'informations dont le citoyen est prêt à se dessaisir, c'est-à-dire à fournir, en renonçant définitivement à exercer le moindre contrôle sur le traitement qu'elles subiront, et sans savoir entre quelles mains elles tomberont. Cette attitude s'explique par le fait que l'information qui semble la plus anodine à première vue peut s'avérer nuisible pour l'intéressé si elle est intégrée à d'autres données » (Rodotà, 1974, 149-150).

Ce choix peut aussi s'expliquer par l'importance croissante qu'allait prendre le concept d'autodétermination informationnelle, préfiguré dans l'arrêt *Mikrozensus*[23] de 1969, et que le Tribunal constitutionnel fédéral d'Allemagne consacrait par l'arrêt *Volkszählungsurteil* de 1983[24], préfiguré par *Mikrozensus*. Selon ce principe, une personne doit en principe avoir donné son consentement pour que des informations à son sujet fassent l'objet d'un traitement.

Le consentement de l'individu, et donc sa capacité à déterminer lui-même, en fonction de ses préférences et de ce qui provoquait (ou non) chez lui un sentiment d'intrusion, allait devenir la pierre angulaire du droit des données personnelles dans l'UE avec l'article 7 de la directive de 1995 puis l'article 8 de la Charte des droits fondamentaux, qui consacrent tous deux le consentement comme le premier des critères de licéité d'un traitement de données personnelles (Rossi, 2019).

Ces choix de mesures concrètes sont cohérents avec une approche libérale de la vie privée, d'ailleurs critiquée par certains pour une diversité de motifs qu'il n'est pas notre objet de discuter ici (Allen, 1999 ; Etzioni, 1999 ; Fuchs, 2011). Cette approche s'est imposée comme une forme de paradigme à ce premier réseau d'acteurs que C. Bennett qualifie de défenseurs de la vie privée (*privacy advocates*) (Bennett, 2008).

Ces choix ont nécessité d'être défendus face à l'idée généralement répandue parmi les décideurs·ses publics·ques de l'époque que l'ordinateur allait accroître la rationalité de la décision publique, notamment en matière économique. Plutôt qu'une tentative de remise en cause frontale de cette idée, la stratégie

23 BVerfG, 16 juillet 1969 - 1 BvL 19/63 (Mikrozensus).

24 BverfG, 15 décembre 1983 - 1 BvR 209/83, 1 BvR 484/83, 1 BvR 420/83, 1 BvR 362/83, 1 BvR 269/83, 1 BvR 440/83 (Volkszählungsurteil).

rhétorique que nous constatons à la lecture des documents d'archives montre que les premières lois de protection des données ont été présentées comme de nature à garantir la confiance des citoyens dans l'informatique. Les deux citations ci-dessous sont des exemples d'illustrations de cette stratégie dont nous trouvons la trace dans les documents d'archives de notre corpus :

> « [...] afin d'éviter tout malentendu, le préambule réaffirme que l'utilisation d'ordinateurs à des fins d'administration publique devrait généralement être considérée comme un progrès. L'objectif de la présente Résolution n'est pas de s'opposer à une telle utilisation mais de la renforcer en posant certaines garanties » (Note explicative accompagnant le projet de résolution sur la protection de la vie privée vis-à-vis des banques de données électroniques dans le secteur public, Conseil de l'Europe, EXP/Prot.Priv. (73) 11, 6).
> « L'objectif fondamental doit être de restaurer la confiance du public à l'égard du traitement des informations retenues par l'administration » (Blekeli, 1974, 16).

Nous verrons que cet argument de la confiance est resté central dans la stratégie argumentative de la coalition des défenseurs de la vie privée, plusieurs décennies plus tard, au moment des négociations sur la proposition de RGPD présentée par la Commission européenne en janvier 2012.

## Défendre une acception large de la notion de donnée personnelle dans le RGPD au nom de la confiance du citoyen européen dans l'informatique

Entre l'adoption de la Convention 108 en 1981 et celle du RGPD en 2016, l'informatique s'est considérablement développée et ses usages massivement répandus. Internet fut ouvert à un usage commercial au milieu des années 1990 (Rebillard, 2012). L'effondrement de la bulle des dotcoms, autour de l'an 2000, a poussé les acteurs et actrices de l'industrie désormais dite du numérique à exploiter à des fins économiques et à très grande échelle les données personnelles des internautes (Cecere, Le Guel et Rochelandet, 2015 ; Fuchs, 2012), contribuant au développement du capitalisme de surveillance contemporain (Zuboff, 2018).

Sur le plan politique et idéologique, le tournant néolibéral (Jobert, 1994) a modifié les arguments en faveur d'une exploitation plus libre d'applications informatiques permettant le traitement à grande échelle de données personnelles. Si l'État est resté un utilisateur central, son rôle s'est recentré sur des fonctions régaliennes. En matière de collecte de données, cela s'est traduit, notamment depuis les attentats du 11 septembre, par le perfectionnement de systèmes de surveillance de masse au nom de la lutte contre le terrorisme (Lyon, 2015). Mais ces activités régaliennes étant partiellement en dehors du champ du droit de l'Union, c'est principalement au nom d'une croissance économique « fondée sur la connaissance », défendue par la Commission depuis la stratégie de Lisbonne[25], que sont justifiées les demandes visant à assouplir le droit de l'Union en matière de données personnelles. Rhétoriquement, nous retrouvons dans plusieurs documents l'idée qu'un « juste équilibre » (*right balance*) devrait être trouvé entre cet intérêt général économique (tel qu'interprété dans un référentiel néolibéral) et le droit à la vie privée, comme illustré par cette situation d'un *position paper* de l'Industry Coalition for Data Protection

> « Nous invitons la Commission à **équilibrer** judicieusement la **protection des droits individuels** avec le fonctionnement du Marché Unique. La capacité de la Société européenne de l'information à **générer de l'innovation et de la croissance** [...] dépend [...] également de l'utilisation de tous les types de données au cœur de l'économie numérique. Des règles trop strictes, statiques et bureaucratiques [...] auront un impact négatif sur l'économie numérique de l'Europe[26] » (*position paper* de l'Industry Coalition for Data Protection de 2011[27], p. 2).

Concrètement, la coalition industrielle favorable à un tel assouplissement a notamment proposé, dès les consultations organisées en 2009 et en 2010-2011 sur une future réforme de la directive de 1995, de distinguer deux types de données : les données personnelles, à « haut risque », touchant directement la sphère de l'intime, et les données dites pseudonymes, qui n'identifiant pas les personnes directement par un nom ou un autre identifiant direct, seraient à plus faible risque. Selon Amazon, il serait ainsi nécessaire d'« inciter les entreprises à investir dans des mesures de protection de la vie privée, comme

25 La stratégie de Lisbonne est disponible en ligne à l'adresse : <https://www.europarl.europa.eu/summits/lis1_fr.htm> (page consultée le 6 août 2022).

26 « We urge the European Commission to balance in a sensible manner the protection of individual rights with the functioning of the Single Market. The ability of the European Information Society to generate innovation and growth, as envisaged in the European Commission's Digital Agenda, depends [...] also on the continued use of all kinds of data that are at the heart of the digital economy. Overly strict, static and bureaucratic [...] rules will have a detrimental impact on Europe's digital economy. » Texte mis en gras par nos soins.

27 Ce document peut être téléchargé en ligne sur le dépôt *Github* du projet Lobbyplag : <https://github.com/lobbyplag/lobbyplag-data/blob/master/raw/lobby-documents/eudataprotect.pdf> (page consultée le 17 mars 2023).

la pseudonymisation, et de les utiliser[28]. » Pour y parvenir, selon l'association EuroISPA, qui fédère les fournisseurs d'accès à Internet en Europe : « une définition claire des données pseudonymes doit être introduite. En outre, dans les cas où les données sont utilisées pour distinguer des utilisateurs, plutôt que pour les identifier, les données ne devraient pas être considérées comme à caractère personnel[29]. » Ainsi, bien que certains, à l'instar de Nihit Goyal *et al.* (2021), insistent sur les divergences d'intérêts entre secteurs de l'industrie, et notamment entre opérateurs de télécommunications (fédérés pour certains au sein d'euroISPA) et les fameux « GAFAM » américains dont fait partie Amazon, ces divergences n'ont pas porté atteinte à la cohérence d'ensemble de la coalition industrielle. La diversité des membres de l'Industry Coalition for Data Protection, allant du e-commerce aux fournisseurs d'accès à Internet en passant par les industries médiatiques, en témoigne[30].

À cette première coalition de cause composée majoritairement de groupes d'intérêts industriels, s'est opposée une coalition favorable au paradigme libéral de la vie privée, regroupant notamment des autorités de protection des données et des associations de défense des consommateurs et des droits fondamentaux. Cette seconde coalition regroupait ainsi celles et ceux que C. Bennett (2008) a appelés les « défenseurs et défenseuses de la vie privée » (« *privacy advocates* »). Comme la première, cette seconde coalition était aussi traversée par des divergences, débats et tensions. Les autorités de protection des données et les associations comme European Digital Rights n'ont pas toujours les mêmes points de vue.

Outre l'analyse des *position papers* de notre corpus et ce qui ressort de la perception des acteurs et actrices de ces coalitions de cause avec lesquels nous

28 Traduit de l'anglais. Texte original : « Businesses should be incentivized to invest in and use privacy-enhancing measures, such as pseudonymization [...]. ». Courrier d'Amazon contenant des propositions d'amendements, disponible en ligne sur le dépôt Github de Lobbyplag : <https://github.com/lobbyplag/lobbyplag-data/raw/master/raw/lobby-documents/amazon_letter_26_1.pdf> (page consultée le 17 mars 2023.

29 Traduit de l'anglais. Texte original : « A clear definition of pseudonymous data should be introduced. Also, in cases where data is used to distinguish between users, rather than identifying them, the data should not be considered as personal. » *Position paper* d'EuroISPA, disponible en ligne sur le dépôt Github de Lobbyplag : <https://github.com/lobbyplag/lobbyplag-data/raw/master/raw/lobby-documents/1212_EuroISPA_contribution.pdf> (page consultée le 17 mars 2023).

30 Voir les signataires du *policy paper* de 2011 cité plus haut, intitulé « Industry Coalition for Data Protection Paper on proposals for a "New EU legal framework on data protection" », disponible en ligne à l'adresse : <https://github.com/lobbyplag/lobbyplag-data/blob/master/raw/lobby-documents/eudataprotect.pdf> (page consultée le 17 mars 2023.

avons effectué des entretiens, une visualisation réalisée avec Gephi en utilisant l'algorithme ForceAtlas 2 (Jacomy *et al.*, 2014) des liens entre amendements et propositions formulées dans des *position papers* de groupes d'intérêts permet aussi de voir que le débat se structure autour du clivage entre deux coalitions rivales :

**Visualisation des coalitions de cause lors de négociations sur le RGPD, faisant apparaître deux regroupements qui se dessinent clairement : coalition des défenseurs et défenseuses de la vie privée en haut, coalition industrielle en bas. D'après : Rossi, 2020, 312.[31]**

31 https://www.theses.fr/2020COMP2549/document.

L'algorithme ForceAtlas2 permet de spatialiser un réseau dans lequel les nœuds qui ont des liens entre eux s'attirent, et les autres s'éloignent. La visualisation ci-dessus montre en orange les député·e·s européen·ne·s signataires de propositions d'amendements, qui sont elles-mêmes liées aux groupes d'intérêts, identifiés en orange, ayant proposé le même amendement (points violets) mot pour mot. La taille des labels est fonction du nombre d'amendements proposés par un·e député·e ou groupe d'intérêt. Les axes ne sont pas signifiants.

Le corpus est constitué des données relatives à des amendements parlementaires reprenant mot pour mot des propositions de groupes d'intérêts, collectées et structurées dans des fichiers au format JSON par Lobbyplag[32], un collectif militant, pendant les négociations sur le RGPD. Il concerne 48 eurodéputé·e·s et 11 groupes d'intérêts pour 357 amendements. L'illustration ci-dessus n'a donc pas force probante en elle-même, car il est impossible de savoir quelle est la proportion de propositions de groupes d'intérêts qui n'ont pas été répertoriées par le collectif Lobbyplag. Mais la proportion d'amendements analysés n'est pas totalement marginale. Elle correspond à un peu moins de 9 % du nombre total d'amendements déposés dans le cadre des négociations sur le RGPD, et elle corrobore nos observations tirées de la lecture de notre corpus de *position papers* et des verbatims d'entretien et nous amène à conclure que les négociations sur le RGPD se sont bel et bien structurées autour de deux pôles : visible en haut, la coalition des défenseurs de la vie privée, et en bas, la coalition industrielle.

Les propositions de cette dernière d'assouplir les règles à appliquer à des données pseudonymes, ont suscité de vives critiques de la part de la coalition rivale des défenseurs et défenseuses de la vie privée, toujours attaché·e·s majoritairement à ce paradigme libéral de la vie privée qui insiste sur l'importance de celle-ci pour l'autonomie individuelle, et s'alarmant de la confusion entretenue dans ces discours entre l'anonymat et le pseudonymat, ce dernier demeurant très vulnérable aux attaques de réidentification (Ohm, 2010) :

> « Si vous regardez le scandale [*Cambridge Analytica*] de Facebook au sujet des élections [...] : cela n'a pas d'importance de savoir que M. Smith, ici, est de 5 River Street dans le Minnesota, est M. Smith de 5 River Street. Mais le problème est la capacité à l'identifier comme quelqu'un de susceptible de recevoir ce message, et d'être manipulable non pas sur la base de qui il est, mais de ce qu'il est. [...] Imaginez si vous pouvez être distingué [*singled out*] et manipulé sans

32 Voir : <https://github.com/lobbyplag/>.

> votre consentement ou à votre insu, et que cela n'était même pas pris en compte dans le droit de la protection des données [*this wasn't even a concept that data protection law would cover*]. Ce serait ridicule[33] » (Entretien avec Joe McNamee, de l'association EDRi, le 9 novembre 2017, cité dans : Rossi, 2020, 552-553).

La tactique déployée en réaction a été d'admettre l'introduction d'une définition de la pseudonymisation dans le RGPD art. 4 (5), tout en définissant la donnée issue de ce processus comme appartenant toujours à l'ensemble des données à caractère personnel, et en en faisant une nouvelle obligation, qui consiste à imposer aux personnes qui traitent des données de chercher à les pseudonymiser au maximum dès lors que c'est possible. Cela a permis de neutraliser la proposition de la coalition industrielle.

Interrogée sur les stratégies argumentatives déployées pour convaincre de leurs propositions, l'une des enquêtés, employée d'une ONG polonaise active au sein d'European Digital Rights (EDRi), une coalition d'associations européennes militant pour les « libertés numériques », insistait sur l'importance de l'argument de la confiance :

> « Habituellement, nous répondons en rappelant que lorsque vous examinez les recherches faites sur le sentiment des gens, vous voyez qu'ils ont le sentiment de ne pas avoir de pouvoir, que [...] les gens ne savent pas ce qui se passe avec leur information, et c'est une question de [...] confiance, principalement de confiance. Il doit y avoir certaines garanties, certaines règles. [...] Et il faut essayer de construire sur la confiance. [...] Si vous arrivez à créer de la confiance, vous obtenez un nouveau modèle économique et vous pouvez être compétitif[34] » (Entretien avec Anna Walkowiak, de la Fondation Panoptykon, citée dans : Rossi, 2020, 360-361)

33 Traduit de l'anglais. Texte original : « If you look at the Facebook scandal around the [...] elections: there is no value in knowing that Mr. Smith here is from 5 River Street, Minnesota, is called Mr. Smith from 5 River Street. But the matter is the ability to identify him as somebody susceptible to receiving this message, and being manipulable on the basis of not who he is, but what he is. [...] Imagine if you could be singled out and manipulated without your consent or knowledge, and if this wasn't even a concept that data protection law would cover. How ridiculous would that be? »

34 Traduit de l'anglais. Texte original : « Usually we respond by reminding them that when you look at some research about how people feel, they feel that they don't have power, that [...] they don't know what's going on with information, and it's a question of [...] trust, mainly trust. There should be some safeguards, some rules. [...] And you're trying to build on trust. [...] If you're able to create trust, you have a new business model, you can be competitive that way. »

Ainsi, comme dans la période allant de 1970 à 1981, la stratégie des défenseurs et défenseuses de la vie privée pour légitimer leurs revendications n'a pas consisté en une attaque frontale à l'égard de l'idée selon laquelle le traitement automatisé de données personnelles était utile économiquement. Ils et elles ont, au contraire, admis dans leurs discours cette utilité, mais cherché (avec succès) à imposer les mesures qu'ils et elles soutenaient comme indispensables à une confiance sans laquelle il ne pourrait y avoir de développement de cette économie de la donnée.

La continuité que nous observons donc dans la définition de la notion de donnée à caractère personnel, comme dans la liste des principes juridiques de la protection des données qui, à l'instar du principe de limitation des finalités, remontent aux propositions formulées par la British Computer Society en 1971 et reprises par le groupe d'experts du Conseil de l'Europe dans la rédaction de deux recommandations adoptées en 1973 et 1974, semble donc au moins en partie s'expliquer par la grande stabilité de la structure argumentative de ce champ de l'action publique qui, depuis le début, s'est mis en place à un niveau européen.

## Étudier la dimension cognitive et la structure argumentative des politiques publiques européennes

Les politiques publiques européennes de la protection des données ont fait l'objet de diverses études (Bennett, 1992 ; Karaboga, 2018 ; Newman, 2008) qui ont notamment permis d'analyser les facteurs de son européanisation, ainsi que le rôle de la coalition des défenseurs de la vie privée (Bennett, 2008). Les travaux dont nous présentons ici les résultats s'est centrée sur un aspect peu étudié jusque-là : la structure argumentative autour de laquelle s'articule le champ de l'action publique et qui en explique les produits (au sens des *policy outputs*).

Une politique publique se compose d'un contenu, ou ensemble de mesures concrètes qui en forment la substance, incluant des décisions plus ou moins coercitives, destinées à un public et visant un objectif (Muller, 2011, 21-22), lequel est dicté par des valeurs et des représentations négociées. Ces actions ainsi que les rivalités entre acteurs cherchant à influer sur leur production s'inscrivent dans un « cadre général de l'action » (Muller, 2011, 21), c'est-à-dire des catégories d'interventions dans lesquelles certains acteurs, actrices et institutions se spécialisent sur un temps long. Nous parlons alors de « secteurs » de l'action publique (Jobert et Muller, 1987), de « champs de l'action publique » (Dubois, 2014) ou encore de « sous-systèmes de politiques publiques » (Sabatier, 1993). Ceux-ci ont la caractéristique de se structurer sur le temps long (Sabatier, 1998).

Les approches cognitives des politiques publiques (Hall, 2015 ; Muller, 2000) s'intéressent aux représentations et aux discours qui façonnent la production de l'action publique. Elles s'intéressent également aux phénomènes de changement et d'apprentissage dans les politiques publiques. Deux théories en particulier ont été retenues : celle de l'*Advocacy Coalition Framework* (ACF) développée par Hank Jenkins-Smith et Paul Sabatier, qui a été adaptée à l'analyse des politiques publiques européennes (Rozbicka, 2013 ; Sabatier, 1998), et celle de Pierre Muller sur l'articulation du « rapport global-sectoriel » (RGS) (Muller, 2011).

Les travaux appliquant l'ACF tendent à montrer que les sous-systèmes de politiques publiques se structurent autour de deux coalitions rivales. C'est le cas autant dans les travaux qui portent sur des politiques publiques nationales (ex : Brown et Stewart, 1993 ; Kübler, 2002 ; Mawhinney, 1993) qu'européennes (ex : Brooks, 2018 ; von Malmborg, 2021). S'il existe un débat sur le point de savoir si les coalitions de cause des politiques publiques européennes se forment de façon stable sur un temps long, ou bien de façon temporaire et opportuniste sur des sujets *ad hoc* (Rozbicka, 2013), nos observations sur le terrain de la protection des données tendent à faire apparaître deux coalitions : une coalition industrielle à la structure relativement récente, et une « *privacy community* » (Bennett, 2008) dont l'ossature remonte aux réseaux d'expert·e·s, défenseurs et défenseuses de la vie privée, constitués à partir des années 1970. Cette dernière coalition comprend notamment des ONG comme EDRi ou la Quadrature du Net, mais aussi des membres d'autorités de protection des données, et un certain nombre de député·e·s européen·ne·s, comme Jan Philipp Albrecht, rapporteur du RGPD.

La première coalition cherchait à remettre en cause le consensus inscrit dans la directive de 1995 dans le sens d'une déréglementation partielle facilitant l'usage de données personnelles, sortant assez largement les données pseudonymes du champ d'application de ces politiques, et reconnaissant une valeur juridique au consentement tacite. Ces revendications allaient dans le sens des intérêts matériels des membres des groupes d'intérêts constituant la majeure partie de cette coalition, tels que ceux-ci les percevaient. Ces discours étaient proches du référentiel global néolibéral, là où les défenseurs de la vie privée sont restés attachés au paradigme libéral (classique) de la vie privée.

L'approche proposée par P. Muller permet de décrire les stratégies par lesquelles un groupe parvient à ancrer ses revendications en les légitimant vis-à-vis d'un discours dominant au sein des élites administratives et politiques d'un État. Elle attire l'attention sur le rapport qu'il y a entre le référentiel sectoriel, c'est-à-dire un paradigme dominant au sein d'un secteur donné, et le référentiel global, « image sociale de toute la société, c'est-à-dire une représentation globale autour de laquelle vont s'ordonner, se hiérarchiser, les différentes représentations sectorielles » (Jobert et Muller, 1987, 65). L'hypothèse de P. Muller (1984), soutenue par ses travaux sur les changements dans la politique agricole française dans la période d'après-guerre, est que si un référentiel sectoriel ne parvient pas à s'articuler au référentiel global, c'est-à-dire à s'exprimer d'une façon qui soit cohérente avec ce dernier, alors les acteurs et actrices qui soutiennent les politiques qui découlent de ce référentiel risquent de perdre le pouvoir face à une coalition proposant un nouveau référentiel sectoriel plus propice à s'articuler au référentiel global. Bien qu'elle ait été pensée pour l'analyse de politiques publiques nationales, nous pouvons mobiliser cette approche, complémentaire à celle de l'ACF, pour l'étude de politiques publiques européennes.

Le paradigme libéral (au sens du libéralisme classique) de la vie privée s'est imposé dès les années 1970 comme le référentiel sectoriel de ce champ de l'action publique. Le tournant néolibéral aurait pu rendre inefficace l'ancienne stratégie d'articulation du référentiel sectoriel favorable à la vie privée au référentiel global sur le terrain. Certains papiers publiés sur le sujet, dont un article de C. Raab et C. Bennett (2020), estiment que nous aurions assisté, entre les années 1990 et l'adoption du RGPD, à un affaiblissement du paradigme de la vie privée. Ils se fondent pour l'affirmer sur l'existence de critiques académiques de ce paradigme, et citent notamment celle qu'a développée Helen Nissenbaum (2010). Or, les éléments empiriques que nous avons précédemment exposés montrent au contraire que les partisans d'une forte protection de la vie privée et des données à caractère personnel ont gardé la même stratégie argumentative fondée non pas sur une tentative de remise en cause frontale du référentiel global, mais sur l'idée que le bénéfice de l'exploitation des données personnelles ne pourra se réaliser qu'à la condition d'une confiance qui serait garantie par un droit robuste des données personnelles.

Ce faisant, ce que nous proposons d'appeler la *structure argumentative* de ce secteur de l'action publique est resté stable et elle a ainsi contribué à stabiliser dans le temps le contenu du droit des données personnelles en Europe malgré les changements d'arènes (passage du Conseil de l'Europe aux institutions communautaires), de technologie (développement massif d'Internet et de ses usages) et de contexte politique (depuis le tournant néolibéral).

Cette structure argumentative n'est pas celle d'un discours d'un·e acteur ou actrice en particulier, au sens où on l'entend généralement (Snoeck Henkemans, 2001). Henri Bergeron, Yves Surel et Jérôme Valluy, s'inspirant de l'ACF, définissent une politique publique sectorielle comme « le produit d'un système de croyances spécifique, issu de la confrontation continue et des compromis successifs passés entre les systèmes de croyances propres à chaque coalition en présence dans un sous-système donné » (Bergeron, Surel et Valluy, 1998, 202-203). La structure argumentative d'un sous-système de politique publique donne alors à voir ce jeu d'arguments et de contre-arguments, fondés sur des référentiels distincts, qui s'échangent de part et d'autre des lignes de clivages entre coalitions de causes rivales. Combinés les uns aux autres, ils se stabilisent selon un rapport de force idéologique, la coalition qui l'emporte dans un sous-système donné parvenant à imposer que son système de convictions devienne le référentiel sectoriel, qui fixe le cadrage du problème public auquel se dédie un sous-système de politique publique, et contribue à en déterminer la production (au sens des *policy outputs*). Étudier cette structure sur le temps long, à l'instar de ce que P. Muller avait fait dans son ouvrage sur les transformations des politiques agricoles françaises à partir de la Libération (Muller, 1984), apporte un éclairage utile sur l'évolution d'un rapport de force entre coalitions rivales au sein d'un secteur de l'action publique, et sur la relation entre cette évolution et celle de l'action publique produite. Cela permet aussi de prendre du recul critique par rapport à l'impression de nouveauté permanente que dégage généralement l'étude du rapport entre droit et technologie.

## Conclusion : retour sur l'étonnante stabilité du paradigme libéral de la vie privée au sein de la structure argumentative des politiques européennes de protection des données

Les approches cognitives des politiques publiques, initialement conçues pour une application à des terrains nationaux, s'adaptent très bien au contexte européen. Elles permettent notamment de suivre l'évolution dans le temps d'une politique publique, y compris lorsque celle-ci se déploie, comme dans le cas étudié dans cet article, en partie successivement et en partie parallèlement au sein d'arènes différentes : OCDE, Conseil de l'Europe, puis CEE et UE. À leur lumière, il est possible de concevoir les politiques publiques comme des secteurs, ou sous-systèmes – les termes changent, mais l'objet qu'ils désignent reste fondamentalement le même – qui sont en définitive des champs au sein desquels évoluent des acteurs et actrices en concurrence pour imposer des mesures concrètes correspondant à leurs systèmes de convictions. Généralement, ces sous-systèmes tendent à se cristalliser autour d'une rivalité centrale entre deux coalitions de cause, dont le clivage se matérialise dans la structure argumentative dessinée par les échanges d'arguments et de contre-arguments visant à les légitimer, cette légitimation passant notamment par des stratégies d'articulation du système de convictions d'une coalition au référentiel global de l'action publique dans une entité politique donnée.

Déjouant les attentes, la structure argumentative du sous-système de politique publique européen de protection des données est restée stable, depuis sa formation au début des années 1970 à la période des négociations sur le Règlement général de protection des données, qui s'est étalée de 2009 – au moment où la Commission annonce son intention de réforme d'une directive antérieure de 1995 – à 2016. Cette structure se caractérise par la prééminence d'un paradigme libéral classique de la vie privée, porté par une même coalition de défenseurs et défenseuses de la vie privée, qui demeure articulé avec succès aux référentiels globaux de l'action publique européenne (du Conseil de l'Europe, puis également des institutions communautaires).

De fait, le contenu du droit produit depuis les années 1970 repose sur la même définition de la notion de donnée à caractère personnel, et sur les mêmes grands principes directeurs, qui correspondent à ce paradigme libéral, ce qui remet en cause le stéréotype selon lequel le droit serait toujours « en retard » sur la technique. Et cette stabilité s'observe malgré les profondes transformations du milieu technique que le droit des données personnelles a vocation à réguler.

Aujourd'hui, une large part des débats actuels sur le droit de l'intelligence artificielle en gestation avec la récente proposition de règlement sur l'IA[35] rappelle ainsi des réflexions que l'on trouvait déjà, par exemple, dans le rap-

35 Proposition de règlement du Parlement européen et du Conseil établissant des règles harmonisées concernant l'intelligence artificielle (Législation sur l'intelligence artificielle) et modifiant certains actes législatifs de l'Union, présentée par la Commission le 21 avril 2021. COM 2021/206/FINAL.

port du Conseil d'État de 1970 en France. Ceci confirme l'intérêt qu'il y a à étudier sur un temps long l'évolution des discours, idéologies et imaginaires qui structurent les clivages, les débats, et les compromis au sein d'un secteur de l'action européenne, y compris de ceux qui ont pour objet d'imposer des règles juridiques et politiques au milieu technique.

## Bibliographie

**Allen, Anita** (1999), « Coercing Privacy », *William and Mary Law Review*, vol. 40 (3), p. 723-757.

**Bennett, Colin J.** (1992), *Regulating privacy: Data protection and public policy in Europe and the United States*, Ithaca, Cornell University Press.

**Bennett, Colin J.** (2008), *The privacy advocates: resisting the spread of surveillance*, Cambridge, Mass., MIT Press.

**Bennett, Colin J. et Raab, Charles D.** (2003), *The Governance of Privacy. Policy Instruments in Global Perspective*, Aldershot, Ashgate.

**Bennett, Colin J. et Raab, Charles D.** (2020), « Revisiting the governance of privacy: Contemporary policy instruments in global perspective », *Regulation & Governance*, vol. 14 (3), p. 447-464.

**Bergeron, Henri, Surel, Yves et Valluy, Jérôme** (1998), « L'*Advocacy Coalition Framework*. Une contribution au renouvellement des études de politiques publiques ? », *Politix*, vol. 11 (41), p. 195-223.

**Blekeli, Ragnar Dag** (1974), « Normes à observer pour le traitement de l'information et les procédures de contrôle », *in Questions d'ordre politique soulevées par la protection des données et des libertés individuelles, principes et perspectives. Compte rendu du séminaire*, Paris, Organisation de Coopération et de Développement Économique (OCDE), p. 75-95.

**Boucher, Philippe** (1974), « SAFARI ou la chasse aux Français », *Le Monde*, 21 mars, p. 9.

**Bowen, Glenn A.** (2009), « Document Analysis as a Qualitative Research Method », *Qualitative Research Journal*, vol. 9 (2), p. 27-40.

**British Computer Society** (1971), « Submission of Evidence to the Committee on Privacy », Londres, British Computer Society.

**British Computer Society** (1972), *Privacy and the computer--steps to practicality: a review of recent work carried out by the Privacy and Public Welfare Committee of the British Computer Society*, Londres, British Computer Society.

**Brooks, Eleanor** (2018), « Using the Advocacy Coalition Framework to understand EU pharmaceutical policy », *European Journal of Public Health*, vol. 28 (supplément 3), p. 11-14.

**Brown, Anthony E. et Stewart, Joseph Jr.** (1993), « Competing Advocacy Coalitions, Policy Evolution and Airline Deregulation », *in* Paul A. Sabatier et Hank C. Jenkins-Smith (dirs), *Policy change and learning: an advocacy coalition approach*, Boulder, Westview Press, p. 83-103.

**Cecere, Grazia, Le Guel, Fabrice et Rochelandet, Fabrice** (2015), « Les modèles d'affaires numériques sont-ils trop indiscrets ? », *Réseaux*, vol. 189 (1), p. 77-101.

**Chambre des représentants des États-Unis d'Amérique** (1966), *Hearings before a Subcommittee of the Committee on Government Operations. House of Representatives, 89th Congress, Second Session*, Washington, U.S. Government Printing Office.

**Conseil d'État** (France) (1970), *Les conséquences du développement de l'informatique sur les libertés publiques et sur les décisions administratives*, Rapport annuel 1969-1970, Paris, Conseil d'État.

**Dubois, Vincent** (2014), « L'État, l'action publique et la sociologie des champs », *Swiss Political Science Review*, vol. 20 (1), p. 25-30.

**Eberlein, Burkard et Newman, Abraham** (2006), « Innovating EU Governance Modes: the rise of incorporated transgovernmental networks », paper presented at the 15th International Conference of Europeanists, 29 March-2 April, Chicago.

**Etzioni, Amitai** (1999), *The Limits of Privacy*, New York, Basic Books.

**Favro, Karine** (2017), « La démarche de compliance ou la mise en œuvre d'une approche inversée », *LEGICOM*, vol. 59, p. 21-28.

**Flichy, Patrice** (2001), *L'imaginaire d'Internet*, Paris, La Découverte.

**Fuchs, Christian** (2011), « Towards an alternative concept of privacy », *Journal of Information, Communication and Ethics in Society*, vol. 9 (4), p. 220-237

**Fuchs, Christian** (dir.) (2012), *Internet and surveillance: the challenges of Web 2.0 and social media*, New York, Routledge.

**Gallouédec Genuys, Françoise et Maisl, Herbert** (1976), *Le secret des fichiers*, Paris, Cujas.

**Gellert, Raphaël** (2016), « We Have Always Managed Risks in Data Protection Law: Understanding the Similarities and Differences between the Rights-Based and the Risk-Based Approaches to Data Protection », *European Data Protection Law Review*, vol. 2 (4), p. 481-492.

**Glancy, Dorothy** (1979), « The Invention of the Right to Privacy », *Arizona Law Review*, vol. 21 (1), p. 1-39.

**Goyal, Nihit, Howlett, Michael et Taeihagh, Araz** (2021), « Why and how does the regulation of emerging technologies occur? Explaining the adoption of the EU General Data Protection Regulation using the multiple streams framework », *Regulation & Governance*, vol. 15 (4), p. 1020-1034.

**Guerrand, Roger-Henri** (1987), « Espaces privés », *in* Philippe Ariès et Georges Duby (dirs), *Histoire de la vie privée. Tome 4 : De la Révolution à la Grande Guerre*, Paris, Seuil, p. 325-411.

**Haas, Ernst B.** (1958), *The Uniting of Europe: Political, Social, and Economic Forces, 1950-1957*, Stanford, Stanford University Press.

**Hall, Peter A.** (2015), « Chapitre 10 / Cognitive Approaches: A French Touch? », *in* Laurie Boussaguet, Sophie Jacquot, Pauline Ravinet et Pierre Muller (dirs), *Une French touch dans l'analyse des politiques publiques ?*, Paris, Presses de Sciences Po, p. 237-262.

**Holvast, Jan** (2013), *De volkstelling van 1971: verslag van de eerste brede maatschappelijke discussie over aantasting van privacy*, Zutphen, Paris.

**Hondius, Frits Willem** (1975), *Emerging data protection in Europe*, Amsterdam, Elsevier.

**Jacomy, Mathieu, Venturini, Tommaso, Heymann, Sebastien et Bastian, Mathieu** (2014), « ForceAtlas2, a Continuous Graph Layout Algorithm for Handy Network Visualization Designed for the Gephi Software », *PLOS ONE*, vol. 9 (6), e98679.

**Jobert, Bruno** (1994), *Le tournant néo-libéral en Europe : idées et recettes dans les pratiques gouvernementales*, Paris, L'Harmattan.

**Jobert, Bruno et Muller, Pierre** (1987), *L'État en action*, Paris, Presses Universitaires de France.

**Karaboga, Murat** (2018), « The emergence and analysis of European data protection regulation », *in* Julia Schwanholz, Todd Graham et Peter-Tobias Stoll (eds), *Managing Democracy in the Digital Age*, Dordrecht, Springer, p. 29-52.

**Kay, Adrian** (2005), « A Critique of the Use of Path Dependency in Policy Studies », *Public Administration*, vol. 83 (3), p. 553-571.

**Kirby, Michael** (1980), « Transborder Data Flows and the "Basic Rules" of Data Privacy », *Stanford Journal of International Law*, vol. 16, p. 27 *sq*.

**Koopmans, Thijmen** (1974), « Interim Report of the Netherlands Government Committee on the Protection of Privacy in Personal Data Registration », EXP/ Prot.Priv. (74) 3, Strasbourg, Conseil de l'Europe, 29 June.

**Kübler, Daniel** (2002), « Les acteurs associatifs dans l'advocacy coalition framework : application aux politiques publiques de lutte contre la drogue en Suisse », *Pyramides. Revue du Centre d'études et de recherches en administration publique*, vol. 6, p. 83-102.

**Laurer, Moritz et Seidl, Timo** (2021), "Regulating the European Data-Driven Economy: A Case Study on the General Data Protection Regulation", *Policy & Internet*, vol. 13 (2), p. 257-277.

**Lynskey, Orla** (2016), « The 'Europeanisation' of Data Protection Law », *Cambridge Yearbook of European Legal Studies*, vol. 19, p. 252-286.

**Lyon, David** (2015), *Surveillance After Snowden*, Cambridge, Polity Press.

**Malmborg, Fredrik von** (2021), « Exploring advocacy coalitions for energy efficiency: Policy change through internal shock and learning in the European Union », *Energy Research & Social Science*, vol. 80, p. 102248.

**Mawhinney, Hanne B.** (1993), « An Advocacy Coalition Approach to Change in Canadian Education », *in* Paul A. Sabatier et Hank C. Jenkins-Smith (dirs), *Policy change and learning: an advocacy coalition approach*, Boulder, Westview Press, p. 59-82.

**Miller, Arthur R.** (1971), *The assault on privacy: computers, data banks, and dossiers*, Ann Arbor, University of Michigan Press.

**Moravcsik, Andrew** (1991), « Negotiating the Single European Act: national interests and conventional statecraft in the European Community », *International Organization*, vol. 45 (1), p. 19-56.

**Muller, Pierre** (1984), *Le Technocrate et le Paysan*, Paris, Éditions ouvrières.

**Muller, Pierre** (2000), « L'analyse cognitive des politiques publiques : vers une sociologie politique de l'action publique », *Revue française de science politique*, vol. 50 (2), p. 189-208.

**Muller, Pierre** (2011), *Les politiques publiques*, Paris, Presses universitaires de France.

**Mumford, Lewis** (1938), *The culture of cities*, New York, Harcourt, Brace and Company.

**Musiani, Francesca** (2015), « Edward Snowden, l'« homme-controverse » de la vie privée sur les réseaux », *Hermès*, vol. 73 (3), p. 209-215.

**Newman, Abraham** (2008), *Protectors of Privacy. Regulating Personal Data in the Global Economy*, Ithaca, Cornell University Press.

**Niblett, G.B.F.** (1971), *Digital Information and the Privacy Problem*, OECD Informatics Studies, volume 2, Paris, OCDE.

**Nissenbaum, Helen F.** (2010), *Privacy in context: technology, policy, and the integrity of social life*, Stanford, Stanford Law Books.

**OCDE** (1974), *Questions d'ordre politique soulevées par la protection des données et des libertés individuelles, principes et perspectives. Compte rendu du séminaire 24-26 juin 1974*, Paris, OCDE.

**Ohm, Paul** (2010), « Broken Promises of Privacy: Responding to the Surprising Failure of Anonymization », *UCLA Law Review*, vol. 57, p. 1701-1777.

**Packard, Vance** (1965), *The Naked Society*, New York, Pocket Books.

**Pierson, Paul** (2000), « Increasing Returns, Path Dependence and the Study of Politics », *The American Political Science Review*, vol. 94 (2), p. 251-267.

**Prost, Antoine** (1987), « Frontières et espaces du privé », *in* Philippe Ariès et Georges Duby (dirs), *Histoire de la vie privée. Tome 5 : De la Première Guerre mondiale à nos jours*, Paris, Seuil, p. 13-154.

**Quelle, Claudia** (2017), « The 'Risk Revolution' in EU Data Protection Law: We Can't Have Our Cake and Eat It, Too », *in* Ronald Leenes, Rosamunde van Brakel, Serge Gutwirth et Paul de Hert (dirs), *Data Protection and Privacy: The Age of Intelligent Machines*, Oxford et Portland, Hart Publishing, p. 33-62.

**Rebillard, Franck** (2012), « La genèse de l'offre commerciale grand public en France (1995-1996) : entre fourniture d'accès à l'Internet et services en ligne "propriétaires" », *Le Temps des médias*, vol. 18 (1), p. 65-75.

**Regan, Priscilla M.** (1995), *Legislating Privacy: Technology, Social Values, and Public Policy*, Chapel Hill, University of North Carolina Press.

**Rodotà, Stefano** (1974), « Protection de la vie privée et contrôle de l'information : deux sujets d'inquiétude croissante pour l'opinion publique », *in Questions d'ordre politique soulevées par la protection des données et des libertés individuelles, principes et perspectives. Compte rendu du séminaire*, Paris, OCDE, p. 149-163.

**Rossi, Agustín** (2018), « How the Snowden Revelations Saved the EU General Data Protection Regulation », *The International Spectator*, vol. 53 (4), p. 95-111.

**Rossi, Julien** (2019), « L'hypothèse de la vie privée des Anciens et des Modernes », *in* Éric George (dir.), *Numérisation de la société et enjeux sociopolitiques. Tome 1. Numérique, communication et culture*, Londres, ISTE, p. 77-88.

**Rossi, Julien** (2020), *Protection des données personnelles et droit à la vie privée : enquête sur la notion controversée de « donnée à caractère personnel »*, Thèse de doctorat, Compiègne, Université de technologie de Compiègne.

**Rozbicka, Patrycja** (2013), « Advocacy coalitions: influencing the policy process in the EU », *Journal of European Public Policy*, vol. 20 (6), p. 838-853.

**Sabatier, Paul A.** (1993), « Policy Change over a Decade or More », *in* Paul A. Sabatier et Hank C. Jenkins-Smith (dirs), *Policy change and learning: an advocacy coalition approach*, Boulder, Westview Press, p. 13-39.

**Sabatier, Paul A.** (1998), « The Advocacy Coalition Framework: revisions and relevance for Europe », *Journal of European Public Policy*, vol. 5 (1), p. 98-130.

**Sénat des États-Unis d'Amérique** (1967), *Computer Privacy: Hearings Before the United States Senate Committee on the Judiciary, Subcommittee on Administrative Practice and Procedure, Ninetieth Congress, First Session, Ninetieth Congress, Second Session, on Mar. 14-15, 1967, Feb. 6, 1968*, Washington, U.S. Government Printing Office.

**Snoeck Henkemans, Francisca** (2001), « Argumentation Structures », *in* Frans H. van Eemeren (dir.), *Crucial Concepts in Argumentation Theory*, Amsterdam, Amsterdam University Press, p. 101-134.

**Turner, Fred** (2008), *From counterculture to cyberculture: Stewart Brand, the Whole Earth Network, and the rise of digital utopianism*, Chicago, University of Chicago Press.

**Vitalis, André** (2010), « France », *in* James B. Rule et Graham Greenleaf (dirs.), *Global privacy protection: the first generation*, Cheltenham, Edward Elgar, p. 107-140.

**Ware, Willis H.** (1967), « Security and Privacy in Computer Systems », Santa Monica, Rand Corporation.

**Warren, Samuel D. et Brandeis, Louis D.** (1890), « The Right to Privacy », *Harvard Law Review*, vol. 4 (5), p. 193-220.

**Westin, Alan F.** (1967), *Privacy and Freedom*, New York, Atheneum.

**Younger, Kenneth** (1972), *Report of the committee on privacy. Chairman: Kenneth Younger. Presented to Parliament by the Secretary of State for the Home Department, the Lord High Chancellor and the Secretary of State for Scotland by command of Her Majesty, July 1972*, Londres, H.M.S.O.

**Zuboff, Shoshana** (2018), *The age of surveillance capitalism: the fight for a human future at the new frontier of power*, New York, PublicAffairs.

**Julien Rossi** est maître de conférences à l'UFR Culture et communication de l'Université Paris 8, chercheur au Centre d'études sur les médias, les technologies et l'internationalisation. Il coordonne le Groupe de travail sur la gouvernance et la régulation d'Internet du GDR Internet, IA et société du CNRS. Ses travaux portent sur les politiques publiques de protection des données à caractère personnel, et sur les rapports entre différents instruments de l'action publique - notamment droit et standards techniques - dans les arènes publiques et privées de la gouvernance d'Internet.

contact@julienrossi.com

POLITIQUE EUROPÉENNE
N° 81 | 2023
Chloé Bérut
p. 86-109

**Strategic Europeanisation**

**A “Context-driven” Approach of the Use of European Instruments in Digital Health Policies**

The aim of this article is to explore how and why EU soft regulations affect member states’ digital health (‘eHealth’) policies. More precisely, this article focuses on changing domestic contexts to demonstrate how difficulties in national digital health initiatives may result in domestic actors using EU instruments with the purpose of moving their agenda forward (‘EU usage’ approach). Actors’ strategies and Europeanisation are analysed through four case studies detailing specific sequences of the development of digital health policies in Ireland and in Austria between 2000 and 2012. The results show that identical eHealth policy entrepreneurs used differently the EU when they were submitted to transformed domestic contexts, highlighting how varying contextual and individual resources affect Europeanisation processes.

**L’européanisation stratégique**

**Contextes nationaux et usages des instruments européens dans la construction des politiques de santé digitale**

L’objectif de cet article est d’explorer les effets des orientations européennes non contraignantes sur le développement des politiques de santé digitale (ou « e-santé ») dans les États membres. Pour ce faire, une perspective centrée sur les contextes nationaux est adoptée, afin d’analyser comment certaines difficultés rencontrées lors de la formulation de politiques nationales de santé digitale peuvent donner lieu à des usages de l’Union européenne dans le travail politique des acteurs nationaux. Les stratégies des acteurs et les processus d’européanisation sont étudiés au travers de quatre études de cas détaillant des séquences du développement des politiques d’e-santé en Autriche et en Irlande entre 2000 et 2012. Les résultats de cette analyse montrent que des entrepreneurs de politique publique similaires réalisent des usages différenciés des instruments européens selon le contexte dans lequel ils évoluent. Ainsi, les ressources contextuelles et individuelles dont les acteurs disposent au niveau national ont un effet important sur les processus d’européanisation.

# Strategic Europeanisation

## A "Context-driven" Approach of the Use of European Instruments in Digital Health Policies

**Chloé Bérut**
Université Grenoble Alpes, CNRS, Sciences Po Grenoble, Pacte

## Introduction

In the area of European Union (EU) digital policy-making, the 2016 General Data Protection Directive has attracted the bulk of scholarly attention (e.g. Goyal *et al.*, 2021; Kalyanpur and Newman, 2019; Schünemann and Windwehr, 2021; Bendiek and Römer, 2018). However, EU integration in the digital sector is not limited to such 'hard' policy instruments. Indeed, the European Commission (EC) has also been initiating various cooperation projects resulting in the adoption of soft European guidelines (Featherstone and Radaelli, 2003; Sabel and Zeitlin, 2008) with the aim of harmonising member states' digital policies. Under which conditions should we expect these soft EU incentives to affect member states' domestic policy-making? To answer this question, this article focuses on the area of digital health ('eHealth'), which refers to the integration of Information and Communication Technologies in healthcare. In this policy sector, 'hard' and 'soft' EU regulations (Terpan, 2015) are concomitant. The continuous coexistence of hard and soft EU regulations was for instance particularly visible during the COVID-19 crisis: while member states' cooperation on contact-tracing application took place in the context of informal discussions resulting in the adoption of soft guidelines by the EC, the Green certificate (which contains COVID-19 vaccination proves) was by contrast adopted in a binding legislation.

At the beginning of EU cooperation on digital health however, EU instruments in eHealth were exclusively 'soft' in nature. In the 2000s, the European Commission published numerous Communications, whose aim was to encourage member states to invest in domestic digital health strategies. For instance, references to eHealth policies can be found in the 1999, the 2002, and the 2005 eEurope strategies.[1] In addition, the European Commission published in 2004 an 'action plan for eHealth', which stated that 'by end 2005, [...] member state[s] [should have] develop[ed] a national or a regional roadmap for eHealth'.[2] Later, in 2010, the health sector was included in a general publication on the EU digital strategy (the 2010 EC Communication 'A digital agenda for Europe').[3] With time, these broad recommendations became more and more practical and took the form of technical guidelines, in order to promote European standards for the exchange of digital data in health, such as those gathered in Electronic Health Records or in ePrescriptions (for a detailed presentation of these instruments, see: Bérut, 2021). However, these guidelines remained non-binding. Therefore, in areas such as digital and eHealth policies, which stand at the crossroads of EU and domestic jurisdictions, accurate accounts of Europeanisation must not only focus on hard law transformations, but also on the domestic effects caused by EU soft modes of governance.

In this view, this paper's aim is to explore the effects of soft Europeanisation on member states' digital health policies. At the domestic level, the investment in eHealth policies generally takes the form of the adoption of national-level policy strategy documents, in which various actions are described in order to develop eHealth programs. These eHealth programs can be considered as the 'building blocks' of eHealth policies, and are contributing to the digitalisation of healthcare practices. Very often, one of the first major building blocks in which governments invest is the creation of Electronic Health Records (EHRs), which contain the medical information of the citizens (illness, vaccinations, specific medical conditions, laboratories results, hospitalisations...). Besides investments in EHRs, eHealth initiatives refer to a multitude of other initiatives, such as digital prescriptions (ePrescriptions), telemedicine, secure mes-

1 Communication of 8 December 1999 on a Commission initiative for the special European Council of Lisbon, 23 and 24 March 2000 - eEurope - An information society for all [COM(1999) 687 final - not published in the Official Journal]; COM (2000) 330: eEurope 2002 - An Information society for all - Draft Action Plan prepared by the European Commission for the European Council in Feira - 19-20 June 2000; Communication from the Commission to the Council, the European Parliament, the Economic and Social Committee and the Committee of the Regions - eEurope 2005: An information society for all - An Action Plan to be presented in view of the Sevilla European Council, 21/22 June 2002 [COM(2002) 263 final].

2 Communication from the Commission to the Council, the European Parliament, the European Economic and Social Committee and the Committee of the Regions - *e-Health - making healthcare better for European citizens: an action plan for a European e-Health Area* {SEC(2004)539} /COM/2004/0356 final / p. 16.

3 Communication from the Commission to the European Parliament, the Council, the European Economic and Social Committee and the Committee of the Regions. *A Digital Agenda for Europe*. COM(2010) 245 final/2.

saging, and lots of other digital services specific to certain types of medical conditions (diabetes, cancer...).

This article explains why these domestic initiatives are likely to be affected by European integration in digital health. To do so, a perspective focused on domestic contexts is adopted, in order to demonstrate how difficulties in national eHealth initiatives may result in domestic actors using EU instruments with the purpose of moving their agenda forward ('EU usage' approach, see Jacquot and Woll, 2004; Woll and Jacquot, 2010).

Digital health policies are a particularly relevant case to study the role of changing domestic contexts in Europeanisation. Indeed, these policies present a high degree of complexity and tend to be developed and implemented over several decades, thus leaving the room for manoeuvre to study political or technical changes and their effects on the strategies of 'policy entrepreneurs'(Kingdon, 2002 ; Mintrom and Norman, 2009) promoting eHealth at the domestic level. In addition, the benefit of eHealth policies is often hard to demonstrate (Black *et al.*, 2011), which makes argumentation and references to external organisations supporting digital health (such as the EU) even more crucial in the political process.

Domestic contexts and Europeanisation are analysed through four case studies. Each of this case is about a specific sequence of the development of digital health policies in Ireland and in Austria, between 2000 and 2012. More precisely, the agenda-setting stage of digital health policies in these two countries is compared (in 2000-2001 in Ireland, and in 2005 in Austria), as well as two subsequent eHealth initiatives resulting from the administrative and political work conducted on these subjects (the publication of the 2004 *Health Information, A national strategy* in Ireland, and the adoption of the ELGA[4] law in Austria in 2012). The distinct advantage of these four case studies is that in Ireland and in Austria, the main policy entrepreneurs for eHealth were embodied by the same individuals, while the context in which these policy entrepreneurs evolved differed. This specific diachronic analysis is particularly fruitful to understand how transformations in contexts and actors' resources are key in the explanation of Europeanisation processes in digital health.

4 ELGA is the German acronym for Electronic Health Record (*Elektronische Gesundheitsakte*).

The first section of this paper underlines the benefits of adopting a context-driven approach to explore Europeanisation processes. Indeed, such a perspective appears to be useful to answer specific questions about the timing of EU usage in the policy process. In addition, this section presents the chosen operationalisation for measuring domestic contexts (Kingdon's Multiple Streams theory) and EU usage, as well as the data used to perform the comparative analysis of the Irish and the Austrian cases. The second section is dedicated to the presentation of the results. It focuses first on a comparative analysis of the agenda-setting of eHealth policies in Ireland and in Austria in the beginning of the 2000s. Then, the second part analyses how changing contexts in these two cases resulted in a differentiated use of the EU at the domestic level. By providing an in-depth account of the adoption of digital health policies, this study sheds light on the (un)favourable domestic contexts policy actors evolve in and on how their use of the EU varies over time.

## A "context-driven" approach of soft Europeanisation processes

### Domestic contexts and Europeanisation

To understand Europeanisation processes, this article analyses the way European instruments are 'incorporated in the logic of domestic (national and subnational) discourse, political structures and public policies' (Radaelli, 2003, p. 30). Instead of focusing on the results of Europeanisation (such as transformation, absorption, retrenchment, or inertia), the aim of this article is to study the mechanisms through which the EU becomes part of domestic policy developments. To do so, the 'usage' approach is particularly useful, as it allows focusing on the 'social practices' of actors using the EU in their domestic political work as a 'set of opportunities, be they institutional, ideological, political, or organizational' (Woll and Jacquot, 2010, 116, see also Jacquot and Woll, 2004, 2003; Graziano *et al.*, 2013). In this approach, Europeanisation is conceptualised as the result of the strategic actions undertaken by domestic actors, which are not to be conceived of as simple 'transmission belts' of European institutions. As stated by Sophie Jacquot and Cornelia Woll (2010, 115), this perspective provides a 'focal point for phenomena' which are also studied in other strands of literature, such as the research on blame avoidance and credit claiming in multi-level politics (Hobolt and Tilley, 2014). However, this latter body of work focuses on specific actors (mainly the media or political decision-makers), and not on the many dimensions of the policy cycle, as it is the case with the EU usage perspective. Despite these variations in scope, these approaches share the idea that the EU may be used strategically by domestic actors in order to promote their own agenda, be it in a positive way (e.g., as a support for an argument) or in a 'negative' style (e.g., using the EU to bring discredit on an initiative, see Reungoat, 2014, see also Graziano *et al.*, 2011).

What is also shared by these various approaches is a 'system-driven' type of explanations for the use of the EU in domestic policy-making. For instance, in the blame-shifting literature, explanations for the use of the EU refer to decision-makers' position in multi-level governance systems, to institutional constraints (Heinkelmann-Wild and Zangl, 2020), or to broad economic variables (Traber, Schoonvelde and Schumacher, 2020). In the EU usage literature, hypotheses draw on relatively systemic and stable variables, such as 'EU membership', the 'relationship to Europe', or the 'policy fit/misfit' with EU standards (Graziano, Jacquot and Palier, 2011, p. 8–9). These kinds of institutional variables are extremely useful for broad comparisons, when one aims for instance to compare the global level of EU usage in a country or in a policy sector. However, these wide-ranging variables are of a limited utility for understanding why the EU may be used at specific moments during the development of a public policy. In this case, the question we seek to answer is why the EU may be (or may be not) considered as a strategic tool in the political strategies of domestic policy actors at different points in time (Boussaguet and Jacquot, 2021 ; Duman and Horvath, 2013). To answer this question, this article focuses on the changing domestic contexts and their effects on Europeanisation and policy entrepreneurs' strategies.

The term 'policy entrepreneurs' has been mainly developed in relation to the agenda-setting stage of public policies (Kingdon, 2002 ; Mintrom and Norman, 2009 ; Petridou and Mintrom, 2021). Policy entrepreneurs are key actors who invest their multi-dimensional resources (time, personal connections, financial resources...) to work towards the adoption of a public policy, very often in the long-run. According to John W. Kingdon (2002), they are crucial actors in the policy process as they are the ones who strategically use the political context in order to put a public policy on the governmental agenda. In Kingdon's 1984 seminal theory of agenda-setting, this 'context' is defined by three elements:

1. The political stream, which gathers the various actors of the policy process (the government, civil servants, political parties, public opinion...) and their power relations
2. The problems stream, in which problems compete in order to attract the attention of decision-makers

3. The solutions (or 'policy') stream, in which specialised 'public policy communities' develop solutions likely to respond to public problems

Then, 'windows of opportunity' (e.g., elections, focusing events, or changes in indicators) must be 'seized' by seasoned policy entrepreneurs in order to 'couple' the three streams and to put the public policy on the political agenda. Thus, to understand policy entrepreneurs' strategies, one must not only look at their individual resources, but also at the context in which they evolve. While Kingdon's work focuses on agenda-setting, it has been acknowledged that this type of decisive strategic action could also appear during the other stages of the policy process (Bérut, 2023 ; Herweg, Zahariadis and Zohlnhöfer, 2017 ; Howlett, McConnell and Perl, 2015). In this article, Kingdon's theory is therefore applied to the policy-making process beyond the agenda-setting stage.

## Identifying and comparing EU usage and domestic contexts

In this study, the individuals labelled as 'policy entrepreneurs' are the ones for whom it is possible to establish a connection between their activities in the policy process and the adoption of a specific policy measure in digital health. Empirically, other policy actors very often refer to policy entrepreneurs as the most 'relevant' person in the field of study. To measure their 'resources', the proxy used was policy entrepreneurs' institutional position, which was particularly appropriate in the cases under study as both policy entrepreneurs were part of the political apparatus of their own country.

Regarding contexts, data about the policy process in Austria and in Ireland were gathered. In a first step, data stemming from the grey literature, policy reports, official policy documents and from the media were used in order to gain a general knowledge of the decision-making process of eHealth policies. Then, semi-structured interviews were conducted with policy actors in order to collect more detailed narratives from those individuals likely to have good oversight of the policy process (Ministries heads of departments, civil servants, decision-makers, stakeholders...).[5] Interview guidelines included questions about the work environment of policy actors, the supporters and the opponents to eHealth policies, and the main issues encountered by digital

5 Interviewees were selected first thanks to a positional method (identifying the relevant actors thanks to their institutional position), completed by a reputational method (asking the interviewees who to get in contact with). Interviews were mainly conducted between 2017 and 2020.

health initiatives at a period of time. The complete list of data can be found in the online appendix attached to this article.

Observations collected through this method were first put in a chronological order using Nvivo, and then classified in the Multiple Streams model thanks to the following testing questions: does the observation provide information about actors and their power relations (i.e. the political stream)? Does the observation provide information about the technical aspect of policy solutions (i.e. the policy stream)? Does the observation provide information about the creation or hierarchisation of policy problems (i.e. the problem stream)?

Then, to detect EU usage, a particular method was applied. First, references to European instruments were identified in the written documents related to the selected policy initiatives (see the online appendix for a list of these documents). Then, this information was completed by interview questions geared towards understanding the role of the EU level in the political work of the interviewed actors (participation to the EU level, significance of EU initiatives for the policy actor, etc.). To ensure proper triangulation (Beach and Pedersen, 2013 ; Zeitlin, Barcevičius and Weishaupt, 2014), statements about use of the EU were always supported by two or more different data sources, as was crucial information regarding policy contexts.

To better understand the role of domestic context in policy entrepreneurs' strategies, this article is based on an original research design, in which two policy entrepreneurs, their agenda (developing eHealth policies at the domestic level), and their relationship with the European level remained unchanged across time, while the domestic contexts in which they evolved differed. The four case studies describe two policy sequences regarding the development of digital health policies in Ireland and in Austria. These two countries can be considered as relatively similar, as EU usage is not expected to be particularly prominent, especially when compared to Eastern European countries (Graziano, Jacquot and Palier, 2011 ; Zeitlin, 2009). The analysis is based on a co-variational approach (Blatter and Haverland, 2012) applied between and within cases. By studying how context variations result in a differentiated use of European instruments, this research design sheds light on the strategic components of the explanation of soft Europeanisation processes.

The following section describes the four case studies. The first part focuses on the national agenda-setting of eHealth policies in Austria and in Ireland. The second part details the changing contexts after these agenda-setting periods. In these four case studies, the main policy entrepreneurs for digital health remained identical, that is Mr. Brunner[6] in Austria, and the Minister for Health Micheál Martin in Ireland.

## EU usage as a tool in challenging domestic contexts: the case of digital health

### Using the EU in the agenda-setting of digital health policies: politics, problems, and solution in Austrian and Irish eHealth policies

In this first empirical section, the agenda-setting of eHealth policies in Austria and in Ireland is compared. If both agenda-setting stages occurred in the beginning of the 2000s, the domestic contexts in which they took place were markedly dissimilar, especially regarding three focal points: the wider reform in which eHealth strategies were embedded (problem stream and window of opportunity), the existence of past failures regarding digital health initiatives (solution stream), and a differentiated political endorsement of eHealth policies in the political stream. These various elements contribute to explaining why policy entrepreneurs' contextual resources differed in the two countries, resulting in a varying use of European instruments in Austria and in Ireland.

In Austria, digital health policies appeared on the political agenda during the general healthcare system reform that took place in 2005.[7] This reform, which profoundly affected the Austrian healthcare system, emerged in the aftermath of a particularly exceptional 'window of opportunity', as a coalition between the conservative centre-right political party ÖVP (*Österreichische Volkspartei*) and the far-right FPÖ (*Freiheitliche Partei Österreichs*) party had come into power for the first time after the 1999 general elections. According to several researchers, this political situation gave the ÖVP-party a wide room for manoeuvre to undertake austerity and neoliberalist measures that were on the background of the political agenda since the 1990s (Moreau, 2012 ; Obinger

6 Real name has been changed.

7 More precisely, eHealth was the main topic of the article 7 of the law that enshrined the reform in the Austrian system (the 2005 Austrian agreement pursuant to Art. 15a B-VG on the organisation and financing of the health care system). The article 7 in itself did not include precise measures to be undertook with regard to digital health, but rather consisted in a list of general objectives to be attained. Therefore, it seems that article 7's main purpose was to provide the Austrian federal government with a legal basis in order to pursue the development of digital health policies.

and Tálos, 2010 ; Österle and Heitzmann, 2009). In this context, the Austrian Minister of Health Maria Rauch-Kallat adopted a healthcare reform which pursued three main objectives: an increased efficiency, a better control and steering of the healthcare system, and a better expense allocation between the various financing organisms.[8]

Therefore, the agenda-setting of eHealth took place in a context in which the problems to be solved were mainly linked to the economic stability of the health system. However, a problem stream dominated by financial issues is not necessarily the best option to promote investments in eHealth technologies. Indeed, even if some savings can be expected from digital health devices implementation in the long-run,[9] eHealth instruments are complex and tend to result essentially in additional expenses in the first years, with no direct return on investment (Black *et al.*, 2011 ; Eden *et al.*, 2018). Despite this, decision-makers may resort to the adoption of eHealth solutions in times of economic constraint, using these technologies to present themselves as 'innovators' in the eyes of public opinion. In these cases, eHealth solutions are framed both as an instrument increasing healthcare quality, and as an economically efficient device, even though their concrete ability to reduce healthcare spending is uncertain at best.

Regarding the solution stream, it is worthy to note that in 2005, previous initiatives regarding the administrative side of digital health (such as the electronic e-card) were not fully implemented and thus were associated with political failure. The Austrian e-card was indeed supposed to be implemented by the newly elected government in 1999. But the political turmoil that took place in the aftermath of these elections did not allow for a full implementation of the e-card. Besides, important technical issues emerged at this time (Interview, member of the 2005-2007 'eHealth initiative' committee, Austria, 15/03/2017). In addition, the digitalisation of economic and social activities was far less advanced than today, and eHealth technologies had never been implemented on such a large-scale. Therefore, the problem and the solution streams appeared to be poorly suited for the agenda-setting of federal initiatives in eHealth, especially given the fact that digital health was not part of

8 Hofmarcher, Maria M. et Monika Riedel (2004). 'The Austrian Health Reform 2005'. *Health Policy Monitor*. Available on: <https://www.ihs.ac.at/publications/lib/oa37.pdf>.

9 Mainly through a better healthcare coordination, a reduction of iatrogenic risks, and an increased efficiency in the patient diagnosis and treatment (in particular in emergency situation or in the case of chronic diseases).

the governmental program adopted in 2003 by the ÖVP-FPÖ coalition. The Minister of Health herself did not express any kind of political involvement regarding eHealth technologies,[10] which meant that eHealth did not benefit from the support of high-ranking political actors in the political stream. In this context, how can we explain the agenda-setting of eHealth at this period?

From 2003 to 2005, Mr. Brunner, the main Austrian eHealth policy entrepreneur who describes himself as a 'veteran in eHealth politics' (Interview, Mr. Brunner, main eHealth policy entrepreneur, Austria, 03/03/2017), was working in the office of the Federal Minister of Health M. Rauch-Kallat, and was therefore in an advantageous institutional position. However, it was the first time that this actor occupied a position within the governmental apparatus, and as such he was a relatively new actor in the Ministry for Health politics. While the context of the three streams did not appear as particularly favourable to eHealth solutions, this policy entrepreneur was able to take advantage of the European level in order to secure the inclusion of digital health in the political agenda. His use of European instruments was particularly visible in the article introducing eHealth in the Austrian healthcare reform.[11] Indeed, the 'harmonisation of domestic approaches with European measures and programs' was listed as the third objective to be pursued by eHealth policies, after the improvement of healthcare quality and the economic potential of these technologies. In addition, the possibility of implementing 'binding' guidelines and projects of a 'supra-regional, national, or European' significance was underlined in the article. The European level was therefore particularly visible in the law, and the interview with Mr. Brunner confirmed the significance of EU initiatives during the agenda-setting process of eHealth:

> '[...] And I think in the overall, in the overall development of the last ten-fifteen years, the European Commission had a very positive role to highlight the necessity of eHealth in the agenda, so you know so I am very grateful that they did that, so at certain junctions, the eHealth, the initiatives of the European Com-

---

10 This statement is supported by the many press releases archives provided by the main Austrian news agency *Austria Press Agentur* (APA). To search this database, I used the keyword 'Rauch Kallat' for the period 2003 to 2005. About a hundred of web pages gathering various press releases from this period were reviewed, and I selected, read, and analysed the press releases in which the words 'healthcare/health reform', 'eHealth', or 'tax reform' were included in the title. These press releases show that the eHealth topic is only rarely discussed by the Minister for Health (except for the *e-cards*), and when it is the case, this subject is not explicitly linked to the 2005 healthcare reform.

11 Article 7 of the 2005 Austrian agreement pursuant to Art. 15a B-VG on the organisation and financing of the health care system.

> mission or the European Union or whoever was in charge, mostly the European Commission, was helpful' (Interview, Mr. Brunner, main policy entrepreneur for eHealth, Austria, 03/03/2017).

The influence of the EU was even more visible in the immediate aftermath of the agenda-setting stage, during the adoption of the 'Austrian eHealth initiative'. This document, which is a roadmap for the development of eHealth in Austria, refers several times to the EU, claiming that Austria was responding to the 'requirements' of the European Commission eHealth action plan.[12]

> 'It was useful at this time because we, we could say that we would not... invent eHealth strategy, we had to say that it was a European interest. [...] and especially it was a challenge for the Ministry of Health to implement eHealth, and it should fit in the European program, in the European surroundings' (Interview, leader of the 2005-2007 'eHealth initiative' committee, 20/02/2017).

Therefore, in an unfavourable context for eHealth, the main Austrian policy entrepreneur for eHealth heavily relied on soft European orientations to boost the legitimacy of his political agenda.

Ireland displays a strongly contrasting case when compared to Austria. The topic of eHealth (or 'health information') was firstly introduced on the political agenda by M. Martin, who became Minister for Health after a cabinet reshuffle in 2000 (window of opportunity). The agenda-setting of eHealth took place in a period of economic growth, during which the government's aim was to increase investments in healthcare, and to transform the system into a 'patient-oriented' organisation:[13] at this time, the problem stream was dominated by quality-of-care issues, rather than financing issues.

> 'The aim of this Government, said the Minister, was to place the patient at the centre of our health services. We wish to see services that accommodate patients, rather than the other way around and improved information must be used to improve care for the patient at the level of their hospital or their general prac-

12 Pfeiffer, Karl P. (2007). *Recommendation for an Austrian e-Health Strategy: An information and communication strategy for a modern Austrian healthcare system.* Vienna: publications from the Austrian eHealth initiative, p. 12, see also p. 5 and p. 39.

13 McDaid David, Miriam Wiley, Anna Maresso and Elias Mossialos (2009). 'Ireland: Health system review'. *Health Systems in Transition*, 11(4), p. 199; see also: Department of Health, 27/11/2000 Press release, available on: <https://health.gov.ie/blog/press-release/minister-martin-addresses-the-national-health-information-conference/> [Archived website].

> titioner. New systems that can allow the rapid transfer of information between hospitals, general practitioners and community services will help to ensure a patient centred focus.'[14]

In this case, digital health technologies were strongly supported by the Minister of Health in the political stream, who, as a policy entrepreneur, benefited from an outstanding position of power. The Minister's focus on health information was meant to result in the adoption of a National Health Information Strategy in 2001. His commitment was particularly visible in the publication of the general plan for the Irish healthcare reform in 2001 entitled *Quality and Fairness. A health system for you.* In this strategy, the publication of the National Health Information Strategy was planned for December 2001, and the administrative work on an Irish Electronic Health Record was scheduled.[15]

Regarding the solution stream, there was not, at this time, any record of past failure in digital health initiatives at the national level. Indeed, before the national strategy, eHealth policies were mainly conducted on a regional-scale basis by the Health Boards who were then the main actors in charge of the delivery of care. Therefore, at this time, it is likely that the use of European instruments did not appear as a political necessity for the Minister of Health who just came into power and had the opportunity to develop his program in a favourable economic and political context.

As a result, no instances of EU usage could be observed in the empirical material collected on this period. Interviews with actors working on eHealth at this time did not reveal any connection between the agenda-setting of eHealth in Ireland and the European Union,[16] even though the European Commission had already published two action plans on digital policies (the 1999 and 2000 eEurope action plans), which both referred to eHealth.

---

14 Department of Health, 27/11/2000 Press release, available on: <https://health.gov.ie/blog/press-release/minister-martin-addresses-the-national-health-information-conference/> [Archived website].

15 Department of Health (2001), *Quality and Fairness, a Health System for you.* Dublin: publications of the Department of Health, p. 48.

16 Interview, Health Boards and Health Service Executive senior official, Ireland, 01/12/2017; Interview, Health Boards and Health Service Executive senior official, Ireland, 08/12/2017; Interview, eHealth policy actor, Health Service Executive, Ireland, 18/12/2017; Written correspondence, National Health Information Strategy coordinator, 2001-2002, 07/2019; Written correspondence, eHealth policy actor, Health Service Executive, 11/2017.

In Ireland and in Austria, eHealth policy entrepreneurs occupied very different positions within the political system (member of the Minister's office versus the Minister himself), which shows that policy entrepreneurs do not always benefit from the same type of institutional resources to influence the policy process. In addition, the Austrian agenda-setting of eHealth took place in a context where the problem stream was dominated by financing issues, which are not necessarily the easiest issues to couple with eHealth solutions. By contrast, the problem stream in Ireland at the beginning of the 2000s was influenced by a strong focus on the quality of care and on the creation of a 'patient-oriented' system, in which eHealth technologies were thought to play a prominent role. Finally, if eHealth technologies were quite new and uncertain in the two countries, the solution stream was not tarnished by a record of digital health failures at the national level in Ireland, as it was the case in Austria.

In this view, the Austrian eHealth policy entrepreneur appeared to be more poorly equipped with domestic contextual and individual resources to advance his political agenda on digital health than the Irish Minister for Health. As a result, instances of EU usage only appeared in the Austrian case, in which the policy entrepreneur was able to present the adoption of an eHealth strategy as a 'requirement' from the European Commission.

## Improving and deteriorating contexts and their effects on the use of the EU

In this section, we will see how changing contexts affected the use of EU instruments in Ireland and Austria, while the main policy entrepreneurs for eHealth remained identical during the two periods studied.

In Austria, in the aftermath of the agenda-setting of eHealth, a specific organisation was created in order to work on the technical basis of the Austrian Electronic Health Record 'ELGA'. The role of this organisation, named 'ELGA GmbH', was to make sure that the ELGA record would be ready to be implemented at the beginning of the 2010s, thus strengthening the legitimacy of these technologies within the solution stream. However, for the ELGA record to be fully operational, the technical readiness of the project was not the only challenge to be addressed. Indeed, various decisions related to the design of the ELGA record needed to be included in a specific law (e.g., the information to be contained in the record, the mandatory aspect of the record, the rules for an ELGA record to be officially open...). In this context, eHealth civil servants worked on the preparation of an ELGA bill which was presented to the Austrian Parliament in the early 2010s. This work took place at a specific time regarding the general developments of the Austrian healthcare system. Indeed, after several years of an ÖVP-FPÖ government, the traditional SPÖ (*Sozialdemokratische Partei Österreich*)-ÖVP coalition (social democrat – conservative centre right) came back into power in 2008 and undertook a national healthcare reform, which was used as a window opportunity for the ELGA law. Contrary to the 2005 healthcare reform, the SPÖ-ÖVP reform was more oriented towards the achievement of an improved quality of care – at least in its official objectives. At this time, the Austrian healthcare system was indeed criticized for its poor performances in terms of life expectancy, especially when compared to other European countries.[17] In addition, the Austrian Social Security institutions had managed to achieve a balanced budget for 2011. Therefore, in 2012, the problem stream in Austrian healthcare was a bit less dominated by financing issues, while quality of care issues were put up front. As a result, the political stream was particularly favourable to digital health, and investments in the ELGA record benefited from a strong political agreement between the Federal State, the Länder (regions), and the Social Security institutions.[18]

> 'From 2012 to 2016, something like that I think, and in these periods the goals were very clear that we needed the healthcare record, that it had to be in place, and there was an agreement between the Ministry of Health, the social insurance companies and the regions, as the main drivers of healthcare in Austria, payers, so to say, regulators and payers, they were agreeing on this goal, and ELGA was really heavy part of that' (Interview, eHealth policy actor, Austria, 09/03/2017)

---

17 Bachner Florian, Julia Bobek, Katharina Habimana, Joy Ladurner, Lena Lepuschütz, Herwig Ostermann, Lukas Rainer, Andrea E. Schmidt, Martin Zuba, Wilm Quentin, and Juliane Winkelmann (2018). 'Austria: Health system review', *Health Systems in Transition*, 20:3, p. 184.

18 See the governmental program of the SPÖ-ÖVP coalition for 2008-2013 and the numerous press releases on this subject published by the social securities institutions, e.g. 10/05/2012 Press release from the Austrian Federation of Social Security institutions, Available on: <https://www.ots.at/presseaussendung/OTS_20120510_OTS0124/e-medikation-wissenschaftliche-evaluierung-des-pilotprojekts-bestaetigt-hohen-patientennutzen>; 04/06/2012 Press release from the Austrian Federation of Social Security institutions, Available on: <https://www.ots.at/presseaussendung/OTS_20120604_OTS0198/hauptverband-mit-dem-elga-gesetz-kann-die-e-medikation-oesterreichweit-umgesezt-werden>.

On the other side, digital health policies had benefited from a particularly consistent support from the Ministry of Health over the years. Indeed, during this whole period, eHealth technologies remained under the jurisdiction of the main Austrian policy entrepreneur Mr. Brunner, who became in 2005 one of the few top-ranking civil servants heading the Federal Ministry of Health, a position he was still holding in 2017 at the time of the study. His institutional resources therefore greatly increased with time. All these elements contributed to creating a favourable domestic context with a view to the adoption of the ELGA law. It is worthy to mention however that the ELGA project still met some oppositions at this time, in particular from the Chamber of doctors who launched a wide awareness campaign in order to convince the Austrian citizens of the risks of such an initiative, with the idea that their personal life would be 'exposed' to the government (Interview, eHealth policy actor, Austria, 09/03/2017).[19] In addition, the law had to be negotiated as part of the usual politics of the SPÖ-ÖVP coalition. Therefore, despite a quite favourable context for the adoption of the ELGA law, some oppositions remained, which could have led the main policy entrepreneur for eHealth to refer again to the European level.

However, when looking at the 2012 ELGA law, the contrast with the article 7 of the 2005 Austrian healthcare reform is striking, as almost no references to the EU could be observed.[20] Furthermore, the EU never appeared as an argument in the many publicly available reports on the parliamentary debates that took place at this time, despite the fact that Mr. Brunner directly intervened as a speaker.[21] Besides, the EU was never mentioned during interviews conducted with actors referring to this specific period of time.[22] This is even more surprising given that the European cooperation was at a turning point, with the adoption of

19 See also: press release from the Austrian Chamber of doctors, published on 02/11/2011, available on: <https://www.ots.at/presseaussendung/OTS_20111102_OTS0037/aerztekammer-informiert-bevoelkerung-ueber-gefahren-durch-elga>.

20 The law nevertheless provides a legal basis regarding to the format of health data, which can be aligned on European standards.

21 See also: Parliamentary correspondence n° 825 from the 24th October 2012, 'The Health Committee gives the ELGA Act the green light', Available on: <https://www.parlament.gv.at/PAKT/PR/JAHR_2012/PK0825/index.shtml>.

22 Interview, Mr Brunner, main Austrian eHealth policy entrepreneur, 08/03/2017; Interview, member of the 2005-2007 'eHealth initiative' committee, 15/03/2017; Interview, member of the Austrian Medical Council, 16/03/2017; Interview, Austrian eHealth policy actor, 09/03/2017.

an article on eHealth in the 2011 Cross-border Directive on Healthcare[23] and the creation of the eHealth network, which gathers national authorities on eHealth, and which happened to be chaired at this time by Mr. Brunner himself. Therefore, nor a lack of EU contacts, neither the disappearance of every form of opposition to Federal digital health initiatives may explain the absence of EU usage at this time. Rather, it seems that the increased individual resources of the main policy entrepreneur for eHealth (institutional position) but also more favourable streams and window of opportunity contributed to decreasing the perceived interest in making use of the European level in this specific context.

The Irish case contrasts greatly with the Austrian one, as the political conditions regarding the adoption of eHealth policies tended to deteriorate after the 2000-2001 agenda-setting stage. Indeed, despite the announcement of the *National Health Information Strategy* in 2001, the effective publication of the strategy was constantly delayed.[24] First, in the solution stream, unforeseen difficulties swiftly appeared when civil servants started to work on this national strategy. Indeed, the development of a health information plan required adopting strong data protection measures, as well as an official health identifier.[25] These questions were deemed too sensitive politically, and would have required a law to be adopted by the Irish Parliament. In addition, the political attention shifted at this time towards a major problem which was the structural reform of the Irish healthcare system, the aim of which was to reduce its fragmentation (merging of the regional Health Boards into a single centralised Health Service Executive – HSE).[26] Therefore, eHealth was slowly discarded from the main political agenda of the Ministry of Health, losing its initial support in the political stream.[27]

---

23 Directive 2011/24/EU of the European Parliament and of the Council of 9 March 2011 on the application of patients' rights in cross-border healthcare.

24 Debates of the Dáil Éireann conducted on the 17/04/2002, vol. 552 n° 1, available on: <https://www.oireachtas.ie/en/debates/debate/dail/2002-04-17/232/>; Debates of the Dáil Éireann conducted on the 27/01/2004, vol. 578 n° 4, available on: <https://www.oireachtas.ie/en/debates/debate/dail/2004-01-27/11/#s13>.

25 Written correspondence, coordinator of the National Health Information Strategy, Department of Health (2001-2002), 07/2019.

26 McDaid David, Miriam Wiley, Anna Maresso and Elias Mossialos (2009). 'Ireland: Health system review'. *Health Systems in Transition*, 11(4), p. 200.

27 Between 2001 and 2003, after the announcement of the preparation of a National Health Information Strategy, no references to eHealth could be found in the Department of Health press releases website. To conduct this research, the following keywords were used: 'information', and 'ICT'. Source: <https://health.gov.ie/publications-research/publications/> [archived website].

In 2004 however, a window of opportunity opened in the political stream, thanks to the rotating Presidency of the Council of the EU. While Ireland became the country holding the Presidency, the Minister for Health decided to use this opportunity to put again eHealth on the top of the agenda. The link between the EU and the domestic level in eHealth is clearly visible in the *National Health Information Strategy* which was eventually adopted in 2004, in particular on page 27 where European initiatives in eHealth (eEurope 2005) are described.[28] Besides, data suggest that the Minister of Health M. Martin was particularly active in including eHealth in the four priorities for health of the EU Presidency:

> 'Ireland will have the opportunity to make major advances on health issues during the Presidency of the European Union this year. The action plan for eEurope 2005 includes the area of eHealth and Ireland has made eHealth one of the key themes of the Irish Presidency. It is against this background that, under Ireland's Presidency of the European Union, I have prioritised the area of eHealth [...]' (Welcoming speech from the Irish Minister for Health M. Martin at the 2004 eHealth conference in Cork)[29]

In May 2004, another event contributed to strengthening the connection between the European and the domestic level in digital health. Indeed, the second EU conference on eHealth took place in Cork, for the first time outside Brussels. The organisation of this conference seems to have played a strong catalysing effect on the policy formulation of the *National Health Information Strategy*:

> 'But he [the Minister for Health] said the only way we are going to get anywhere is if we have a national action plan for everything. [...]. So, in 2004, he went to a conference in Cork. And he made an announcement that he was going to have a national eHealth strategy. And then he asked us to write it. [...] It was when the European Commission came to Cork with the eHealth conference. That is why he wanted it for them' (Interview, EU eHealth projects manager, Southern Health Board and Health Service Executive, 04/12/2017).

The document *Health Information. A National Strategy* was finally published in July 2004, just two months after this event. Therefore, while the agenda-setting

28 Department of Health (2004). *Health Information. A National Strategy*. Dublin: publications of the Department of Health, p. 27.

29 Speech available on the conference website: <http://www.ehealthconference2004.ie/> [archived website].

of eHealth occurred in an exclusively domestic frame in 2000, three years of delays and difficulties in the making of the strategy seemed to have turned the EU into a strategic instrument to revive the Irish eHealth national strategy. Table 1 (below) summarises the contextual changes that occurred in the two countries as well as their associated instances of EU usage.

**Table 1: Main changes in domestic contexts and use of the EU in Ireland and in Austria, 2000-2012**

| Country and main policy entrepreneur | 1st case: context and EU usage | 2nd case: context and EU usage |
|---|---|---|
| Austria<br>Main policy entrepreneur: Mr. Brunner | **2005 agenda-setting of eHealth**<br>**Politics: Poor political endorsement**<br>**Problems: Economic constraint**<br>**Solutions: Record of past failures** | **2010-2012 adoption of the ELGA Law**<br>**Politics: Wide political agreement on eHealth**<br>**Problems: Increasing the quality of care**<br>**Solutions: Technical maturation of the ELGA project over several years** |
| | **EU usage** | **No EU usage** |
| **Ireland**<br>**Main policy entrepreneur:**<br>The Minister for Health | 2000 agenda-setting of eHealth<br>Politics: Strong political endorsement by the Minister of Health<br>Problems: Increasing the quality of care<br>Solutions: No record of past failures at the national level | 2004 publication of the national health information strategy<br>Politics: Shift in the political attention<br>Problems: Reducing the fragmentation of the health system<br>Solutions: Significant policy-making issues in eHealth |
| | No EU usage | **EU usage** |

## Conclusion

The aim of this article was to contribute to the literature on Europeanisation by showing how a context-driven approach could provide new insights on the use of the EU by policy entrepreneurs. In this view, the specific research design presented in this paper highlighted how identical eHealth policy entrepreneurs made different use of the EU when they were submitted to changing domestic contexts. In Austria, while the EU level was firstly used as an opportunity to support the inclusion of eHealth in the 2005 healthcare reform, it appeared merely as irrelevant five years later, during the adoption of the ELGA law. This can be explained by variations in the context of domestic policy-making, but also by the increased institutional resources of the main policy entrepreneur for eHealth. By contrast, if the domestic context seemed favourable enough to put eHealth high on the Irish Minister for Health agenda in 2000, three years of delays and difficulties in the making of the National Health Information Strategy turned the EU into an opportunity to revive the eHealth policy process at the domestic level. Therefore, changing contexts and resources of policy actors seem to have a strong effect on the use of EU instruments.

These results show that a refined understanding of Europeanisation requires an in-depth analysis of actors' strategies and of the way these strategies evolve during the policy process. Indeed, more 'stable' variables such as the 'goodness of fit' hypothesis (Börzel and Risse, 2003), or the degree of policy entrepreneurs 'socialisation' to the EU level (Zeitlin, Barcevičius and Weishaupt, 2014) were not sufficient to understand variations in EU usage: both policy entrepreneurs had a similar relationship to the EU during the periods analysed, and EU guidelines in eHealth were generally too broad in the 2000s to generate any form of misfit mechanism (Bergeron, 2005). Therefore, what created opportunities for EU usage was not a discrepancy between European and domestic models, but rather the difficulties encountered by policy entrepreneurs in the making of domestic eHealth strategies. *In fine*, the thorough analysis of public policies 'trajectories' (Bezes and Palier, 2018) appeared to be a crucial element to understand the differentiated use of the EU over time.

However, if this study provides a good overview of the strategies implemented by policy entrepreneurs at a micro-level of analysis, it falls short of bringing to light the broader systemic type of explanations likely to account for Europeanisation processes. Indeed, EU usage does not occur in institutional voids, but should be understood in the wider context of domestic transformations and their connection with EU integration. In this view, the context-driven approach developed in this article is to be considered as an additional element to Europeanisation theories, in which a more complete framework should also incorporate broader variables likely to affect policy entrepreneurs' strategies, such as the global relationship of a country to the EU level (Graziano, Jacquot and Palier, 2011) or the long-term transformation of policy actors' cognitive frames induced by European integration (Borrás and Jacobsson, 2004 ; Kröger, 2009 ; Sabel and Zeitlin, 2008 ; Trubek and Trubek, 2005).

## Bibliography

**Beach, Derek and Pedersen, Rasmus Brun** (2013), *Process-Tracing Methods: Foundations and Guidelines*, Ann Arbor, The University of Michigan Press.

**Bendiek, Annegret and Römer, Magnus** (2018), "Externalizing Europe: the global effects of European data protection", *Digital Policy, Regulation and Governance*, vol. 21 (1), p. 32-43.

**Bergeron, Henri** (2005), "Europeanization of Drug Policies: From Objective Convergence to Mutual Agreement", *in* Monika Steffen (ed.), *Health Governance in Europe: Issues, Challenges, and Theories*, Londres, Routledge, p. 174-187.

**Bérut, Chloé** (2021), "The European Union as an opportunity: structures and uses of European soft law in French, Austrian and Irish eHealth policies", *West European Politics*, vol. 44 (1), p. 155-175.

**Bérut, Chloé** (2023), "The chemical framework: Exploring Europeanisation in French, Austrian, and Irish eHealth policy processes", *Governance*, vol. 36 (4), p. 1063-1081.

**Bezes, Philippe and Palier, Bruno** (2018), "Le concept de trajectoire de réformes", *Revue française de science politique*, vol. 68 (6), p. 1083-1112.

**Black, Ashly D., Car, Josip, Pagliari, Claudia, Anandan, Chantelle, Cresswell, Kathrin, Bokun, Tomislav, McKinstry, Brian, Procter, Rob, Majeed, Azeem and Sheikh, Aziz** (2011), "The impact of eHealth on the quality and safety of health care: a systematic overview", *PLoS medicine*, vol. 8 (1), p. e1000387.

**Blatter, Joachim and Haverland, Markus** (2012), *Designing Case Studies: Explanatory Approaches in Small-N Research*, Basingstoke, Palgrave Macmillan.

**Borrás, Susana and Jacobsson, Kerstin** (2004), "The open method of co-ordination and new governance patterns in the EU", *Journal of European Public Policy*, vol. 11 (2), p. 185-208.

**Börzel, Tanja and Risse, Thomas** (2003), "Conceptualizing the domestic impact of Europe", *in* Kevin Featherstone and Claudio M. Radaelli (eds), *The Politics of Europeanization*, Oxford, Oxford University Press, p. 57-80.

**Boussaguet, Laurie and Jacquot, Sophie** (2021), "Os usos do supranacional : dinâmicas de integraçao regional e o papel dos atores nacionais", *in* Patrick Hassenteufel and Osmany Porto de Oliveira (eds), *Sociologia Política da Ação Pública: Teorias, abordagens e conceitos*, Brasilia, Escola Nacional de Administração Pública (Enap), p. 335-352.

**Duman, Anil and Horvath, Anna** (2013), "Traditional Familialism Served with EU Gravy. Reconciliation Policies in Hungary", *Politique européenne*, vol. 40 (2), p. 22-43.

**Eden, Rebekah, Burton-Jones, Andrew, Scott, Ian, Staib, Andrew and Sullivan, Clair** (2018), "Effects of eHealth on hospital practice: synthesis of the current literature", *Australian Health Review*, vol. 42 (5), p. 568-578.

**Featherstone, Kevin and Radaelli, Claudio Maria** (eds) (2003), *The Politics of Europeanization*, Oxford, Oxford University Press.

**Goyal, Nihit, Howlett, Michael and Taeihagh, Araz** (2021), "Why and how does the regulation of emerging technologies occur? Explaining the adoption of the EU General Data Protection Regulation using the multiple streams framework," *Regulation & Governance*, vol. 15 (4), p. 1020-1034.

**Graziano, Paolo R., Jacquot, Sophie and Palier, Bruno** (eds) (2011), *The EU and the Domestic Politics of Welfare State Reforms: Europa, Europae*, Basingstoke, Palgrave Macmillan.

**Graziano, Paolo R., Jacquot, Sophie and Palier, Bruno** (2013), "Usages et européanisation", *Politique européenne*, vol. 40 (2), p. 94-118.

**Heinkelmann-Wild, Tim and Zangl, Bernhard** (2020), "Multilevel blame games: Blame-shifting in the European Union", *Governance*, vol. 33 (4), p. 953-969.

**Herweg, Nicole, Zahariadis, Nikolaos and Zohlnhöfer, Reimut** (2017), "The Multiple Streams Framework: Foundations, Refinements, and Empirical Applications", *in* Christopher M. Weible and Paul A. Sabatier (eds), *Theories of the policy process*, New York, Routledge, p. 17-53.

**Hobolt, Sara B. and Tilley, James** (2014), *Blaming Europe? Responsibility Without Accountability in the European Union*, Oxford, Oxford University Press.

**Howlett, Michael, McConnell, Allan and Perl, Anthony** (2015), "Streams and stages: Reconciling Kingdon and policy process theory", *European Journal of Political Research*, vol. 54 (3), p. 419-434.

**Jacquot, Sophie and Woll, Cornelia** (2003), "Usage of European Integration – Europeanisation from a Sociological Perspective", *European Integration online Papers (EIoP)*, vol. 7 (12), p. 1-18.

**Jacquot, Sophie and Woll, Cornelia** (2004), *Les usages de l'Europe : Acteurs et transformations européennes*, Paris, L'Harmattan.

**Kalyanpur, Nikhil and Newman, Abraham L.** (2019), "The MNC-Coalition Paradox: Issue Salience, Foreign Firms and the General Data Protection Regulation", *Journal of Common Market Studies*, vol. 57 (3), p. 448-467.

**Kingdon, John W.** (2002), *Agendas, Alternatives, and Public Policies*, 2nd edition, New York, Pearson Education Editions.

**Kröger, Sandra** (ed.) (2009), *What we have learnt: Advances, pitfalls and remaining questions in OMC research*, European Integration online Papers (EIoP), Special Issue 1, vol .13.

**Mintrom, Michael and Norman, Phillipa** (2009), "Policy Entrepreneurship and Policy Change", *Policy Studies Journal*, vol. 37 (4), p. 649-667.

**Moreau, Patrick** (2012), *De Jörg Haider à Heinz-Christian Strache : l'extrême droite autrichienne à l'assaut du pouvoir*, Paris, Éditions du Cerf.

**Obinger, Herbert and Tálos, Emmerich** (2010), "Janus-Faced Developments in a Prototypical Bismarckian Welfare State: Welfare Reforms in Austria since the 1970's", *in* Bruno Palier (ed.), *A Long Goodbye to Bismarck? The politics of Welfare Reform in continental Europe*, Amsterdam, Amsterdam University Press, p. 101-128.

**Österle, August and Heitzmann, Karin** (2009), "Welfare state development in Austria: strong traditions meet new challenges", *in* Klaus Schubert, Simon Hegelich and Ursula Bazant (eds), *The Handbook of European Welfare Systems*, New York, Routledge, p. 31-48.

**Petridou, Evangelia and Mintrom, Michael** (2021), "A Research Agenda for the Study of Policy Entrepreneurs", *Policy Studies Journal*, vol. 49 (4), p. 943-967.

**Radaelli, Claudio M.** (2003), "The Europeanization of Public Policy," *in* Kevin Featherstone and Claudio M. Radaelli (eds), *The Politics of Europeanization*, Oxford and New York, Oxford University Press, p. 27-56.

**Reungoat, Emmanuelle** (2014), "Mobiliser l'Europe dans la compétition nationale", *Politique européenne*, vol. 43 (1), p. 120-162.

**Sabel, Charles F. and Zeitlin, Jonathan** (2008), "Learning from Difference: The New Architecture of Experimentalist Governance in the EU", *European Law Journal*, vol. 14 (3), p. 271-327.

**Schünemann, Wolf Jürgen and Windwehr, Jana** (2021), "Towards a 'gold standard for the world'? The European General Data Protection Regulation between supranational and national norm entrepreneurship", *Journal of European Integration*, vol. 43 (7), p. 859-874.

**Terpan, Fabien** (2015), "Soft Law in the European Union—The Changing Nature of EU Law", *European Law Journal*, vol. 21 (1), p. 68-96.

**Traber, Denise, Schoonvelde, Martijn and Schumacher, Gijs** (2020), "Errors have been made, others will be blamed: Issue engagement and blame shifting in prime minister speeches during the economic crisis in Europe", *European Journal of Political Research*, vol. 59 (1), p. 45-67.

**Trubek, David M. and Trubek, Louise G.** (2005), "Hard and soft law in the Construction of Social Europe: the Role of the Open Method of Co-ordination", *European Law Journal*, vol. 11 (3), p. 343-364.

**Woll, Cornelia and Jacquot, Sophie** (2010), "Using Europe: Strategic action in multi-level politics", *Comparative European Politics*, vol. 8, p. 110-126.

**Zeitlin, Jonathan** (2009), "The Open Method of Coordination and reform of national social and employment policies: influences, mechanisms, effects", *in* Martin Heidenreich and Jonathan Zeitlin (eds), *Changing European Employment and Welfare Regimes: The Influence of the Open Method of Coordination on National Reforms*, Londres, Routledge, p. 214-245.

**Zeitlin, Jonathan, Barcevičius, Egidijus and Weishaupt, Timo J.** (2014), "Institutional Design and National Influence of EU Social Policy Coordination: Advancing a Contradictory Debate", *in* Egidijus Barcevičius, Timo J. Weishaupt and Jonathan Zeitlin (eds), *Assessing the Open Method of Coordination. Institutional Design and National Influence of EU Social Policy Coordination*, Houndmills, Palgrave Macmillan, p. 16-39.

**Chloé Bérut** est postdoctorante et chercheuse en science politique associée au laboratoire Pacte. Ses recherches portent sur la formulation des politiques de santé, en particulier dans un contexte européen. Ses travaux abordent ainsi un ensemble divers de thématiques : santé digitale, inégalités de santé, démocratie en santé. Son identifiant ORCID est 0000-0001-8436-8926.

chloe.berut@unive.it

**POLITIQUE EUROPÉENNE**
**N° 81 | 2023**
Samuel Cipers, Trisha Meyer
p. 110-141

## "Free Speech is Not Free Reach"

### How Platforms Self-regulate Misinformation, Political Ads and Election Campaigns

This article analyses social media platform measures on online political campaigning. We track Meta, Google, Twitter and TikTok and assess which types of content and moderation they prioritise and the frequency of (self-reported) interventions taken in the fight against (political/electoral) mis/disinformation. We analyse their policies on online political campaigning and compare these approaches against the goals set out in the renewed EU Code of Practice on Disinformation and the proposed EU Regulation on the Transparency and Targeting of Political Advertising. We argue that platform policies and practices merit analytical scrutiny as they determine the boundaries of acceptable speech online, and that measures to regulate political speech should be understood as more than addressing political advertising alone, but also should include policies regarding political accounts and mis/disinformation.

## « La liberté d'expression n'est pas la liberté d'accès »

### Comment les plateformes s'autorégulent pour traiter la désinformation, les publicités politiques et les campagnes électorales

Cet article analyse les mesures prises par les plateformes de réseaux sociaux en matière de campagnes politiques en ligne. Sont étudiées Meta, Google, Twitter et TikTok, et notamment les types de contenu et de modération que ces plateformes privilégient ainsi que la fréquence des interventions (auto-déclarées) dans la lutte contre la désinformation (politique/électorale). Nous analysons leurs politiques en matière de campagnes politiques en ligne et comparons ces approches aux objectifs définis dans le nouveau code de conduite de l'UE sur la désinformation et la proposition de règlement de l'UE sur la transparence et le ciblage de la publicité politique. Cet article montre que les politiques et pratiques des plateformes méritent une analyse détaillée car elles déterminent les limites de ce qu'est un discours acceptable en ligne. De plus, ces cas illustrent que les mesures visant à réglementer les discours politiques ne doivent pas se limiter aux seules publicités, mais doivent également inclure les politiques visant les comptes politiques et la désinformation.

# "Free Speech is Not Free Reach"
## How Platforms Self-regulate Misinformation, Political Ads and Election Campaigns

**Samuel Cipers**
Vrije Universiteit Brussel

**Trisha Meyer**
Vrije Universiteit Brussel/United Nations University CRIS in Bruges

## Introduction

The regulation of online political campaigning is a contentious topic. Social media platforms are wary of moderating overtly, and in some ways political content is the ultimate form of free expression, as it can serve to keep those in authority in check (Kaye, 2019). Yet as evidenced in the de-platformisation of US President Trump after the Capitol Riots in 2021, social media platforms are not and cannot be mere 'soapboxes' in the public square. Their algorithms, advertising rules, content moderation and account policies shape what is acceptable form and speech (Gillespie, 2018; Gorwa, Binns and Katzenbach, 2020; Meyer and Alaphilippe, 2021). In Europe, in preparation of the European Parliament elections in 2024, the European Democratic Action Plan (European Commission, 2020a), the renewed Code of Practice on Disinformation (European Commission, 2022) and the proposed EU Regulation on the Transparency and Targeting of Political Advertising (European Commission, 2021) propose next steps in regulating the role of platforms in online political advertising and election campaigns. From ensuring transparency in political advertising to combatting online disinformation, academic and policy views differ, however, on what successful policies should look like, especially with regard to effective measures to curb practices in advertising, while safeguarding democratic rights, and the need for change in the profit-making business model of platforms (Helberger, Dobber and de Vreese, 2021; Tworek, 2019; Zuboff, 2019).

In the aforementioned context, this article seeks to inform the academic and policy debate with data on existing platform measures on online political campaigning. We highlight the results of our monitoring exercise of platform

POLITIQUE EUROPÉENNE N° 81 | 2023

policies and (self-reported) practices to combat (health and political) mis/disinformation over the past two years.[1] While we acknowledge the distinction between misinformation and disinformation, which has tended to focus on the intent of spreading false information (Wardle and Derakhshan, 2017; European Commission, 2018a), in this article we use the terms interchangeably. This is because, as we explain in our analysis below, platforms generally do not base their action on intent (which admittedly can be difficult to measure) but rather on potential to cause harm.

We track four different social media companies (Meta, Google, Twitter and TikTok), analysing which types of content and account moderation they prioritise, and the frequency and timing of new interventions taken in the fight against disinformation. In particular, we note Meta and Google's emphasis on advertising policies as a means to tackle disinformation, which is relatively underlit in evaluations of the functioning of the Code of Practice on Disinformation. Differences between approaches of the companies, trends and shortcomings will also be discussed. In addition, we analyse companies' policies on online political campaigning, which overlap with, yet are distinct from disinformation measures, and make initial comparisons of these approaches against the priorities and goals set out in the renewed Code of Practice on Disinformation and the proposed EU Regulation on the Transparency and Targeting of Political Advertising. We find profound differences between platform community guidelines on political advertisements. In the current regulatory patchwork on misinformation, political advertising and election campaigns, rules for old (media) and new (platform) 'speech intermediaries' differ significantly, resulting in uncoordinated ad hoc platform approaches and leeway for abusive and manipulative use by politicians and political parties.

This research covers a time period of twenty-eight months from January 2020 through April 2022 (the duration of the European Commission COVID-19 disinformation monitoring programme - see more in methodology). This means that the takeover of Twitter by Elon Musk and associated changes to the platform's policies are not included in the analysis. These changes could mark a significant departure from the approach Twitter has taken and result

1 We define 'health dis/misinformation' as dis/misinformation on diseases, available treatments and/or research on diseases. 'Political misinformation' is defined as dis/misinformation on political institutions, parties, politicians, elections and/or other events pertaining to the political level. Although fitting in a larger monitoring project on dis/misinformation, this article focuses on political dis/misinformation only.

in the company coming back on previous decisions in favour of a less strict approach to online misinformation, political advertisements and other content.

Data sources for our analysis comprise the platforms' disinformation monitoring reports to the European Commission, company blog posts and press releases, as well as community guidelines and terms of service.

The article contributes to the special issue and academic literature on digital governance with insights on social media platforms' audited self-regulatory role in determining the limits of acceptable political speech, in a context of ongoing and upcoming EU platform legislation. By analysing the role, strategies, success and ultimately the power of these actors in the policy-making process, we broaden our understanding of EU policy-making in the digital domain. We outline two core arguments: first - platform policies and practices merit analytical scrutiny as they determine the boundaries of acceptable speech online, and second - measures to regulate political speech should be understood as going beyond political advertising alone, but also include policies regarding political accounts and political/election misinformation.

## Background

Platforms receive ample criticism for (perceived) mistakes in moderating political speech. For instance, accusations of being politically biased towards the liberal left were often heard during the Trump presidency. At the same time misuse by malicious actors, in particular Russian interference in elections, was investigated. While recognising that the many-to-many character of communication on platforms is different from traditional media, platforms form an online public sphere (Papacharissi, 2002 – at times even an anti-public sphere, Davis, 2020), in addition to traditional media, in which political ideas and criticism can be deliberated (Häussler, 2017). They are not neutral 'platforms' (soapboxes, stages) from which audiences can be addressed. They amplify and reduce voices, sometimes through paid advertising, sometimes through algorithms. As a result, and due to the importance of political exchange for the well-functioning of democratic societies, online political campaigning (disinformation and advertising) has been under heightened scrutiny in recent years.[2]

2 The ability to harness platforms or oblige them to stifle political opposition through repressive legislative action should also be noted (Bontcheva et al., 2020), but mostly falls beyond the scope of this article.

Platforms draw the boundaries of acceptable speech and behaviour through their community guidelines and terms of service. Their policies are shaped by what constitutes freedom of opinion and expression in international human rights law, select necessary restrictions to freedom of expression, but also platform corporate priorities and consultation with external stakeholders. As many academics and advocates (such as O'Regan, 2018; Tworek, 2019; Posetti et al., 2020; HateAid, 2021; Meyer and Pershan, 2022) have argued, much leeway and room for improvement remains which leads, amongst other problems, to the underperformance of platforms to provide for victims of speech violations. Indeed, through their community guidelines, terms of service, platforms determine which content is illegal and/or potentially harmful and can therefore justifiably be removed from their services, thereby creating their own normative understanding of misinformation and freedom of expression (Gillespie, 2018; Kettemann, 2020). Through legislative initiatives to address online political advertising, curb the spread of mis/disinformation and regulate political advertising, policy makers recognise the *de facto* role of platforms as "speech police" in the online context (Kaye, 2019) and seek to create transparency and give a sense of responsibility (in best cases) – or criminalise in worst (UN Special Rapporteur on the Promotion and Protection of the Right to Freedom of Opinion and Expression, 2018; 2019).

Within the European Union, relevant policy frameworks are the European Democracy Action Plan (European Commission, 2020a), the (revised) Code of Practice on Disinformation (European Commission, 2022) and the proposed Regulation on the Transparency and Targeting of Political Advertising (European Commission, 2021b), alongside broader measures to regulate new technologies, such as the recent Digital Services, Markets, and Data Governance Acts. The European Democracy Action Plan "sets out measures to promote free and fair elections, strengthen media freedom and counter disinformation" (European Commission, 2020b). In order to achieve this goal, amongst others, the European Commission proposed updates to its Code of Practice on Disinformation and legislation on political advertising. The strengthened Code of Practice on Disinformation was signed in June 2022 and revises the 2018 Code of Practice, in which online platforms and advertisers committed through co-regulatory measures to tackle disinformation on their services. The 2022 Code of Practice also includes signatories from civil society, fact checking and research communities. It covers commitments on demonetisation, transparency of political advertising, integrity of services, empowerment of users, fact-checkers and researchers, setting up of a transparency centre and taskforce, and a strengthened monitoring framework (European Commission, 2022a). The Code of Practice complements the Digital Services Act and the proposed Regulation on Transparency and Targeting of Political Advertising. The proposal for a Regulation on the Transparency and Targeting of Political Advertising (European Commission, 2021) builds on the European elections package published in 2018 and aims to prepare for the upcoming EU elections. The regulation would lay transparency obligations for all actors involved in political advertising and limit use of targeting and amplification techniques for political purposes.

Our analysis considers the operation of social media companies as audited self-regulation (as defined by Marsden et al., 2020), especially for the studied time period (2020-2022). With the strengthened Code of Practice and the Digital Services Act, in particular the risk assessment and auditing requirements for very large online platforms, this is now shifting towards a co-regulatory model (Stockmann, 2023). This article seeks to cover both the policies *and* practices of these platforms on online political campaigning, albeit for the latter in a self-reported manner. There are discrepancies between what the rules say, how platforms act upon these rules and whether either corresponds to what is reported. Through our triangulation of sources, we seek to minimise the difference between what platforms report and their rules say. We could not verify the self-reported practices, as this was beyond human sources and data access available in this project. Nonetheless the systematic monitoring of platform measures (policies and self-reported practices) provides a baseline, and perhaps even a good one, against which we can start evaluating and comparing platforms' performance (Gorwa, 2019; Haggart and Iglesias Keller, 2021).

Further, this article covers platform measures on disinformation *and* political advertising, because of the European Democracy Action Plan's scope and our understanding of platform moderation and political speech online. As explained further in the following section (methodology), platforms not only moderate content, but also accounts and advertising. Each of these is relevant when considering online political speech. For instance, despite acknowledging the potential harm caused, Facebook does not allow third party fact-checking of posts of political leaders, motivated by their understanding that these practices could impede freedom of opinion and expression (Harrison, 2021). More positively, labelling the accounts of politicians and political parties allows users to understand the role and function of this person on the platform better and to nuance the content and information provided.[3] While neither example is content moderation in a strict sense, both influence the availability of political speech on platforms.

## Methodology

Our analysis tracks four social media companies: Meta/Facebook (including Instagram), Google (including YouTube), TikTok and Twitter. We grouped different platforms (Facebook and Instagram, Google and YouTube) as this is how these companies have structured their reports to the EU, but indicated differences between platforms of the same companies when we noticed discrepancies. This results in a picture of how companies behind social media platforms have their own distinct approaches to combating misinformation. We ask *how platforms self-regulate online political campaigning through their policies and practices related to (political) misinformation and advertising,* and make an initial comparison of *how these policies and practices relate to EU's expectations in this field.* Our dual focus on (political) misinformation and advertising provides insight into the companies' disposition on regulating misinformation, as well as their general inclination towards online political speech. We can also compare how advertising policies differ in relation to community guidelines and terms of service for regular users on their platforms.[4]

### Data selection

Our first data source for mapping social media companies' measures on online political campaigning was their reports submitted to the European Commission under the COVID-19 disinformation monitoring programme, as part of

---

3 In this article we assume that politicians, political parties or pressure groups are present on social media to advocate, convince and gain followers of their political aims - in other words, are active on platforms for professional, not personal, reasons.

4 It should be noted that anyone, not only politicians and political parties, can easily set up an advertising campaign on social media platforms. In a political context, this means that partisans and (biased) individuals can make their own advertising campaigns to spread ideological messages, support for candidates or counter political adversaries.

the initial Code of Practice on Disinformation.[5] We archived and analysed all available reports from January 2020 through April 2022. The first report covers the start of the pandemic through July 2020, subsequent monthly reports followed from August 2020 through June 2021, and from July 2021 through April 2022 bimonthly reports were published. Reporting stopped when the renewed Code of Practice was agreed upon in June 2022. In these reports, initial signatories to the Code of Practice on Disinformation provide details on their measures to counter misinformation on their platform, including misinformation surrounding the COVID-19 virus, during political events and elections, and other misinformation. These reports provide longitudinal insight into measures taken on the different platforms, which allows us to compare different platforms and to contextualise certain measures in time.

*Blogs and reports* published by the platforms on their own websites formed the second data source. Examining these self-published blogs and other reports (for instance, information available through help desks and transparency centres) complemented the monitoring reports and helped in assessing the narrative and focus of the different companies. It allowed us to research if platforms envision different approaches to online political campaigning, as well as uncover their disposition towards EU policy. Finally, we also examined the social media platforms' *community guidelines and terms of service* related to (political) misinformation and advertising, as the third data source.[6] These reveal in detail how offenders of their policies are addressed on the platforms, what constitutes (political) misinformation and advertising in the eyes of the platforms, and whether there are any exceptions to these policies.

The time period covered for the blogs and other reports follow the monitoring programme (January 2020 through April 2022), while insights gained from consulting the community guidelines and terms of service were limited to the versions published during the period of research (summer 2022). We captured changes to the policies through the other two data sources but acknowledge

5 These reports are available at the following web address (and are archived on the researchers' local drive): European Commission (2022). *COVID-19 disinformation monitoring programme*. Last consulted 20 March 2023, <https://digital-strategy.ec.europa.eu/en/policies/covid-19-disinformation-monitoring>.

6 See Annex B for a full overview of platform blogs, community guidelines and terms of service.

that this 'snapshot' constitutes a possible limitation in our work.[7] The table below shows an overview of the amount of monitoring reports, blogs, community guidelines, community standards and terms of service we analysed per social media company. The amount of measures recorded will be further outlined in the next section.

**Table 1: Types and amount of data sources analysed from January 2020 through April 2022**

| Companies | # Measures recorded | # Reports analysed | # Blogs analysed | # Guidelines analysed |
|---|---|---|---|---|
| Meta | 108 | 17 | 2 | 4 |
| Google | 56 | 17 | 2 | 8 |
| TikTok | 38 | 17 | 1 | 1 |
| Twitter | 107 | 17 | 1 | 3 |
| **Total** | **309** | **68** | **6** | **16** |

## Taxonomy of platform measures

We analysed the three data sources and coded all measures related to health, political and general misinformation (see Annex A)[8], according to a taxonomy of curatorial responses, developed by Meyer in Bontcheva et al. (2020). The reports submitted to the European Commission formed the primary data source, the other two data sources served to verify and complete the coding.

The taxonomy divides platform curation/moderation into multiple categories, related to content (flagging, blocking, limiting, prioritising), account, advertising, research and user-empowerment. In addition, in this article, we specified whether measures are primarily 'reactive/active' (blocking, limiting, account-specific), 'preventative/passive' (user-empowerment, research-specific) or 'mixed' (flagging, prioritising, ad-specific) in nature. In Table 2,

7 The Platform Governance Archive (HIIG, 2023) could present researchers in the future with a solution to this limitation. So far, this archive covers the following platforms: Facebook & Instagram (Meta), Twitter, and YouTube (Google).

8 See screenshot of monitoring efforts in Annex A. Our categorisation necessarily omits the weight of each initiative. Some measures might have a large impact on users and the spread of misinformation, while others will be marginal. We included health misinformation as the monitoring programme has a focus on COVID-19 and this allows for comparison across misinformation types.

we provide definitions for each curatorial category, along with an example from our dataset.

Each identified measure was coded mutually exclusively into one category, by either co-author and reviewed by the other. Cases of doubt or disagreement were discussed and resolved during project meetings. When a measure spanned multiple categories, we either split the measure into multiple entries or we discussed the primary objective and coded accordingly, to avoid double-counting and obtain a crisp data set.

The next two sections present the findings of this coding exercise, highlighting trends, similarities and differences in how the platforms tackle misinformation, followed by insights per platform on political advertising and election campaigning. The discussion and conclusion provide overarching reflections on the state of (self-)regulation and respond to the research question.

**Table 2: Curatorial categories of platform content moderation**

| Category | Definition | Examples |
|---|---|---|
| **Reactive/ Active Measures** | *Reactive measures are designed to deal with misinformation already circulating on the platform and therefore constitute a reaction to past events or current problems.* | |
| Blocking/ Removing of Content | When the platform denies access to certain accounts or shared content by either removing content or the account all together. | Meta/Facebook January 2020: "Announces policy on removal of misleading manipulated media."[9] |
| Limiting/ Demoting of Content | When the platform takes measures to limit the reach of accounts or content. | TikTok October 2020: "reduces discoverability, limits distribution and adds a banner pointing viewers to election guide on content with unverifiable claims." |
| Account-Specific | Platform measures against 'coordinated inauthentic behaviour' and change to account policies to fight misinformation. Blocking, limiting and flagging of accounts fall under previous categories. | Twitter September 2020: "implements account security requirements for high-profile political accounts." |

9 Meta/Facebook January 2020: the report from which this example was taken is entitled, "Facebook response to the European Commission Communication on Covid-19 Disinformation. Report for January to July 2020". For the sake of readability, we only mention the platform and reported date in the table. All examples can be found in Annex A.

| Category | Definition | Examples |
|---|---|---|
| **Mixed Measures** | *Mixed measures tackle misinformation already circulating on the platform, as well as providing actions that increase the resilience of the platform and its users against misinformation* | |
| Flagging/ Labelling of Content | When the platform issues a label to certain accounts or content allowing the users to quickly identify and contextualise content. | Google December 2020: "Google News Initiative provides an additional $1.5M to fund the creation of a COVID-19 Vaccine Media Hub and support new fact-checking research." |
| Prioritising/ Amplifying of Content | When the platform combats the spread of misinformation by promoting verified and accurate information. | TikTok September 2020: "launches elections safety center (in-app guide), accessible from Discover page and on US elections related search results." |
| Ad-Specific | When platforms use ad policies to prevent the spread of misinformation or use ads to promote verified and official information. | Google July 2021: "5M USD in AdGrants made available to governments and organizations in South-East Asia to promote trusted sources and vaccination campaigns." |
| **Preventative/ Passive Measures** | *Preventative measures are designed to limit (the influence of) misinformation campaigns and narratives in the future; in other words, measures to increase the resilience of users against misinformation or decrease the prevalence of false information on the platform in the future.* | |
| User-Specific | Measures that seek to inform the users and provide them with tools to tailor their social media experience to their liking. | Meta/Facebook March 2021: "launched digital media literacy campaign: educate people on how to spot false news online related to vaccine misinformation." |
| Review/ Research-specific | Measures that provide researchers and journalists with the necessary tools to gain insight on the spread of misinformation, the use of ads and to examine the measures taken by the platform. | Google September 2020: "makes available a dataset of search trends for researchers to study the link between symptom-related searches and the spread of COVID-19." |
| **Other** | *All other measures aimed at curbing misinformation on the platforms.* | |
| | Measures that do not fit in a specific category, can nonetheless be included in the above category groups. For instance, providing grants to third parties (not research/ journalism) cannot easily be categorised but fits in the category group 'preventative measures'. | Twitter March 2020: "holds planning exercises to prepare US Elections." |

## Platform measures against misinformation in general

In total, we analysed 309 measures reported by Meta (108), Google (56), TikTok (38) and Twitter (107) to the European Commission. An overview of all measures reported by the studied platforms can be found in Figures 1-3.

**Figure 1: Platform measures to fight misinformation per curatorial category (summary)**

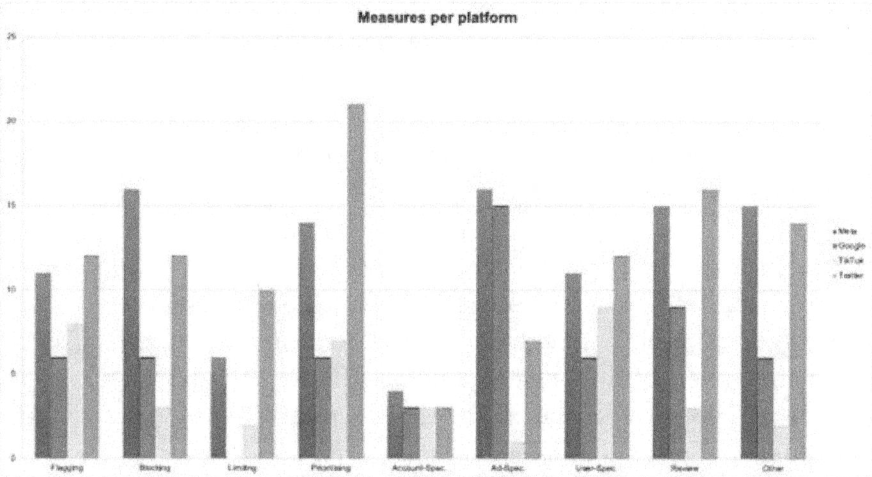


Platforms distinguish harmful from non-harmful misinformation instead of the distinction found in most literature between mis- and disinformation. Whereas the academic and policy distinction focuses on *intent*, the distinction employed by the platforms focuses on the *potential harm* of the message. In function of the seriousness or gravity of possible outcomes, platforms decide whether to intervene. 'Serious' misinformation contains the following across all studied platforms: misinformation with the risk of causing physical harm and misinformation that undermines public trust/interferes in political institutions (supplemented by other 'red lines' specific to each platform).

As an example, in Meta's words:

> *"We remove misinformation where it is likely to directly contribute to the risk of imminent physical harm. We also remove content that is likely to directly contribute to interference with the functioning of political processes and certain highly deceptive manipulated media. In determining what constitutes misinformation in these categories, we partner with independent experts who possess knowledge*

*and expertise to assess the truth of the content and whether it is likely to directly contribute to the risk of imminent harm. This includes, for instance, partnering with human rights organisations with a presence on the ground in a country to determine the truth of a rumour about civil conflict, and partnering with health organisations during the global COVID-19 pandemic. For all other misinformation, we focus on reducing its prevalence or creating an environment that fosters a productive dialogue*" (Meta, n.d.-a).

Both Meta (Facebook and Instagram) and Google (Google Search and YouTube) prioritise their *ad-specific measures* to combat misinformation. Indeed, demonetising misinformation is a crucial step in diminishing the prevalence of misinformation online. Groups or individuals that spread disinformation often have financial objectives. Blocking their ambitions therefore results in less misinformation campaigns being initiated.

Twitter focusses most on *promoting or amplifying content* to offset popular misinformation. The objective of amplifying verified and/or official content is to inform the users before they fall prone to misinformation campaigns and connect users to authoritative sources. It is not only preventative, but additionally informs users and in some cases helps governments to reach their constituents with verified and additional information. Twitter also provides the most measures on *sharing information and data* with researchers and journalists.

Meta and Twitter report the most *reactive measures* to counter misinformation in our sample with a clear lead on TikTok and Google. In the case of Google, it is noteworthy that most of their reactive measures are taken on the YouTube video sharing platform and therefore do not apply to the search engine or other Google services. TikTok reported the least amount of reactive measures.

Our research covers two and a half years of reporting (2020-2022). However, it is clear that reported efforts to combat misinformation have stagnated (across all platforms) since the second half of 2021 (see Figure 2).[10] A possible explanation for this plateau in reported efforts could be a perceived decrease in the danger of the "disinfodemic" (a flood of disinformation related to the coronavirus, infection rates, treatments and vaccinations, Posetti and Bontcheva, 2020)

10 From July 2021 onwards, the platforms shifted from a monthly report schedule to a bimonthly report schedule. The researchers note that the start of the bimonthly report schedule coincided with diminished reported efforts from the platforms.

as pandemic measures eased and vaccination rates rose. We also suspect that EU policy (and reporting) efforts shifted to interinstitutional negotiations on the Digital Services and Markets Acts and multistakeholder revisions of the Code of Practice on Disinformation.

**Figure 2: Platform measures to fight misinformation over time (summary)**

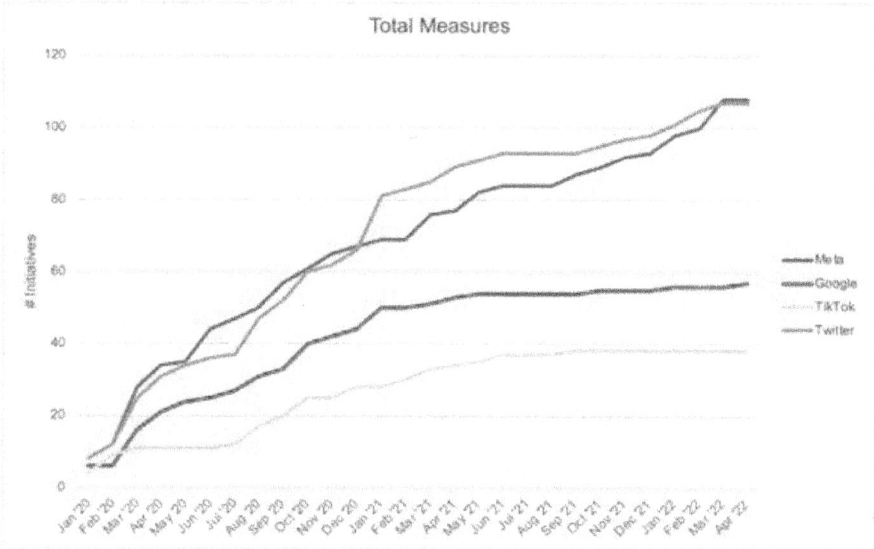


Finally, the following graph (Figure 3) compares how the studied companies approach political misinformation and more general misinformation differently. The y values show how many measures of the given type of misinformation have been reported to the European Commission per platform. The x values are determined by the difference between the number of reactive measures taken and the number of preventative measures taken (Difference $N_{Reac}$ - $N_{Prev}$) per kind of misinformation, per platform. It is important to note that the 'mixed' and 'other' measures are included in the y values (total amount of measures per type of misinformation); however, they are not included in the x values. This graph clearly shows the companies' bias to tackle political misinformation more reactively, whereas general misinformation is combatted with preventative measures. The more measures a company reported the more pronounced this difference becomes.

**Figure 3: Preventative vs reactive responses to misinformation**

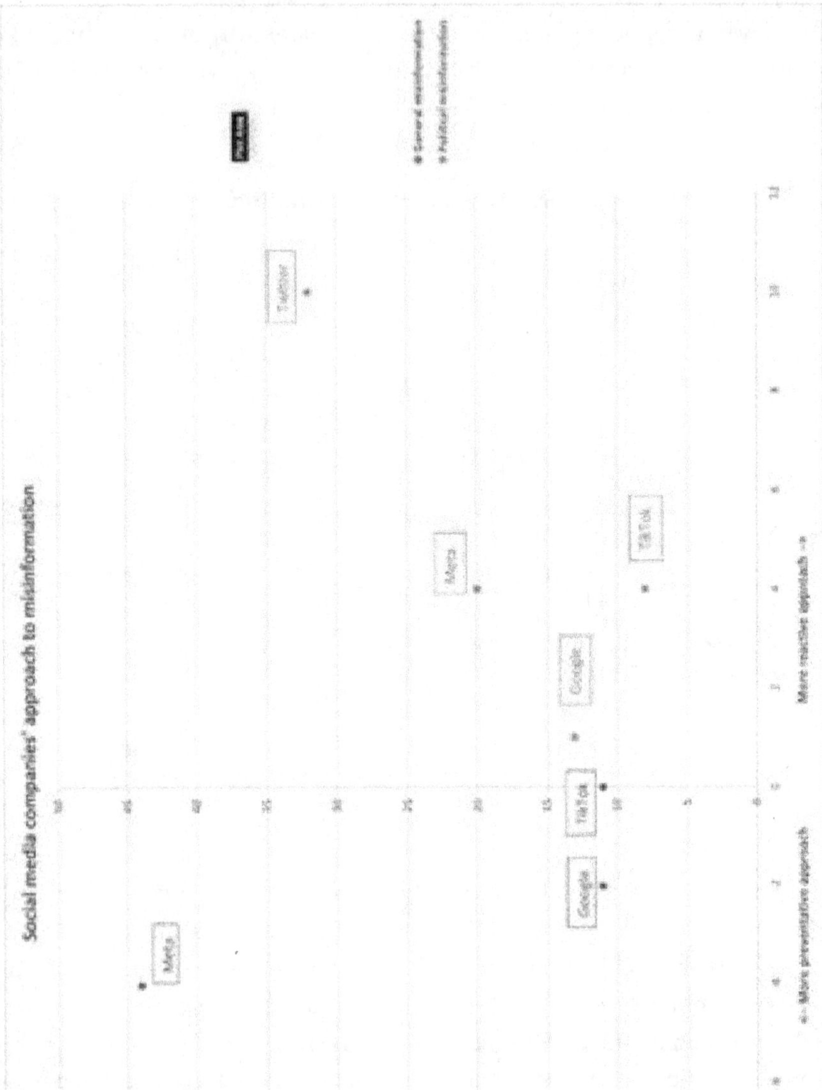

## Platform measures on political advertising and election campaigns

Misinformation and manipulative behaviour that seek to influence voter preferences, interfere in political processes or institutions, or undermine public trust in these institutions have taken a central place in policy makers' minds since the 2016 US presidential elections and the Brexit referendum.[11] Platforms' role has been under scrutiny as well. In this chapter, we look at the measures that social media companies have reported to the European Commission considering political and electoral misinformation.

### Meta

Considering the investigations launched against Facebook for its role in the Cambridge Analytica scandal, the 2016 US elections and UK referendum, Meta has improved their policies and guidelines concerning electoral/political misinformation.

The vast majority of Meta measures to regulate political campaigning are *ad-specific* (see Figure 4). Meta launched its Ad Library in 2018 (Culliford, 2022). Advertisers are verified and each ad is accompanied by information stating which individual or organisation has funded the ad, how much money was invested on ads ("elections ads spending tracker"), and where the sponsor of the ad is located. Users not only benefit from this transparency, but also can choose to turn off political (and social issue) ads or to consult CrowdTangle for an overview of the posts shared by candidates in selected countries. This Ad Library is open to all users, is available worldwide and contains not only political ads, but all ads spread on the platform.

In its policies, Meta also highlights how their technologies can help users vote and works together with third-party fact-checkers, labels misinformation and demotes or deletes content containing misinformation concerning political advertising or electoral misinformation (Meta, n.d.-c). It is noteworthy that these policies do not apply to the content of politicians themselves (Harrison, 2021).

11 Researchers at the Oxford Internet Institute (Bradshaw et al., 2021) found that politicians and political parties in sixty-one countries, among which fourteen EU and NATO countries, made use of "computational propaganda" (which includes online disinformation and manipulation efforts) in 2020.

Furthermore, Meta aims to reduce the prevalence of misinformation on its platform by limiting sharing options or encouraging the reader to reflect on content they have read or want to share. They specify that content containing misinformation that imposes physical harm, interferes with the functioning of political processes (and certain highly deceptive media) is deleted from the platform (Meta, n.d.-a). At the same time Meta commits to respect the voice of its users and will allow content that potentially goes against their standards *"if it's newsworthy and if keeping it visible is in the public interest. We only do this after conducting a thorough review that weighs the public interest against the risk of harm. We look to international human rights standards"* (Meta, 2022). However, as demonstrated with US presidential elections, platform reactivity to events leaves to be desired.

In terms of account level activity, Meta actively labels accounts of governments and politicians. Meta also applies limits and bans to accounts for policy violations more broadly, although it is not as thorough (as we will read below) as Google and Twitter's graduated response policies. In light of foreign interference and coordinated inauthentic behaviour, Meta improved its capabilities to identify and remove malicious accounts and the content they spread.

**Figure 4: Meta's measures to combat misinformation per curational category**

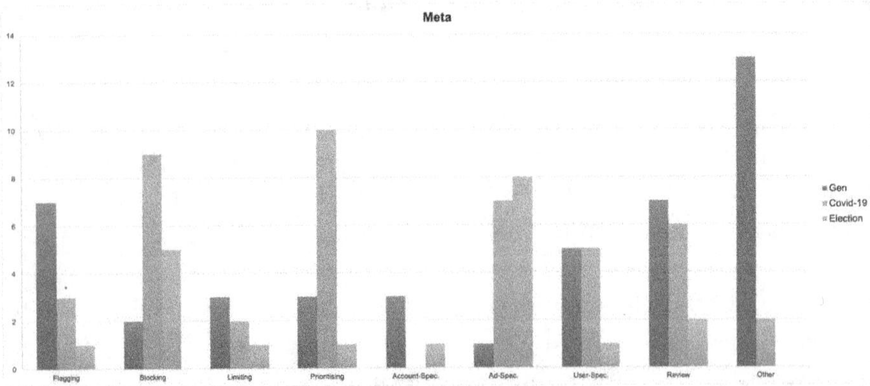


## Google

Similar to Meta, Google focuses primarily on political *ad policies* to combat electoral disinformation (see Figure 5). All political advertisements published since March 2019 are accessible (Google, n.d.-b). Google requires advertisers in the EU to be verified, limits ways in which advertisers can focus on and reach parts of the population and deletes advertisements that violate their policies.

Remarkably, the definition of political advertisement differs depending on the region. For example, ads run by a "referendum campaign group" are seen as "election ads" in the EU, but no such mention is made in the definition of US election ads. This can indicate that Google takes specific political structures and realities into account or limits its involvement in combating misinformation in function of the local law (Google, n.d.-c). Google does indicate paid for advertisements of political actors with a "paid for by" disclosure (Google, n.d.-c).

YouTube (as part of the Google ecosystem) is clear in its "elections misinformation policies", details the forms of content that could be penalised (voter suppression, candidate eligibility, incitement to interfere with democratic processes, distribution of hacked materials, and election integrity) and provides examples. Going against these policies on YouTube constitutes one strike, three strikes ban the user from the platform (Google, n.d.-d).

**Figure 5: Google's measures to combat misinformation per curational category**

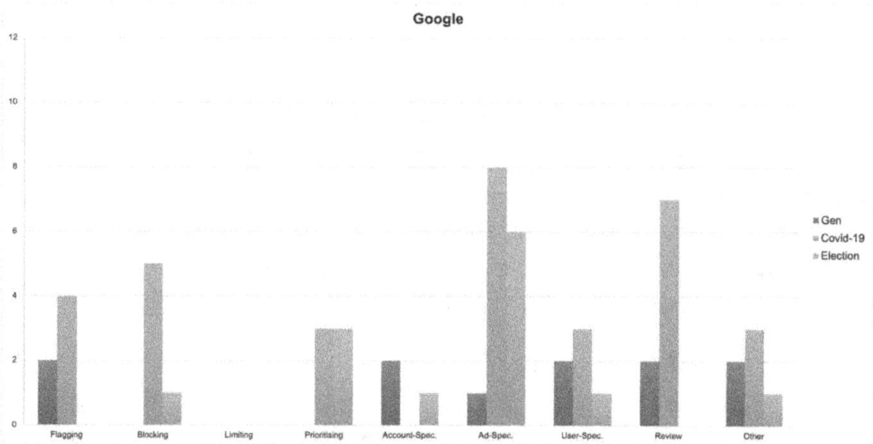


## Twitter

Twitter defines its policies concerning elections a little differently than other platforms, as they have a "civic integrity policy" (Twitter, n.d.-b). Not only do these include political elections, but also censuses, major referenda, and ballot initiatives. Twitter defines four categories of misleading content and behaviour in the aforementioned civic integrity policy: misleading information about how to participate (1) or outcomes (2), suppression and intimidation (3) and false or misleading affiliation (4).

On the platform itself, our data shows that Twitter prefers to combat electoral misinformation mainly by *prioritising authoritative information*, providing tools for researchers and journalists and by blocking offenders (see Figure 6).

Twitter and Meta have in common that their disinformation policies are accompanied by an emphasis on their efforts to provide space for free speech. Twitter provides examples of what users *can* do without repercussions. Interestingly, this includes "inaccurate statements about an elected or appointed official, candidate, or political party" and "organic content that is polarising, biased, hyperpartisan, or contains controversial viewpoints expressed about elections or politics" amongst others (Twitter, n.d.-b).

Twitter also provides account labels on government and state-affiliated accounts. However, concerning political advertisements, in contrast to Meta and Google, Twitter takes a radical approach: political content cannot be promoted.

Twitter also uses a strike system for policy violations, yet differs from YouTube in that some offences grant one strike, but Twitter can attribute more strikes than one to a single offence (depending on the seriousness of the offence). After five strikes (compared to YouTube's three strikes), a user's account is permanently suspended. This allows Twitter to be more flexible in their reactive approaches to disinformation and allows more serious offences to be punished more severely (Twitter, n.d.-b).

**Figure 6: Twitter's measures to combat misinformation per curational category**

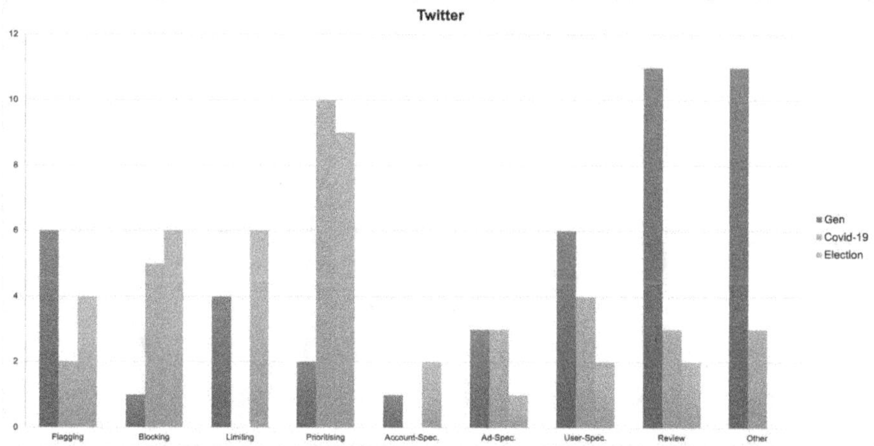


## TikTok

TikTok, much in the same vein as Twitter, does not allow paid political ads to be run on their platform. In their community guidelines (TikTok, n.d.), TikTok reassures its reader that *"[w]e will remove misinformation that causes significant harm to individuals, our community, or the larger public regardless of intent. Significant harm includes serious physical injury, illness, or death; severe psychological trauma; large-scale property damage, and the undermining of public trust in civic institutions and processes such as governments, elections, and scientific bodies"* (TikTok, n.d.). However, TikTok does not provide examples or further define what sort of content would constitute the undermining of public trust.

Aside from this commitment, TikTok aims to connect its users with authoritative information and labels content from accounts belonging to governments, politicians, and political parties (TikTok, 2022, see Figure 7).[12] There is no mention of a strike system or other repercussions that precede the banning of content or accounts.

12 This commitment was found in a resource specific to the US midterm elections of 2022.

**Figure 7: TikTok's measures to combat misinformation per curational category**

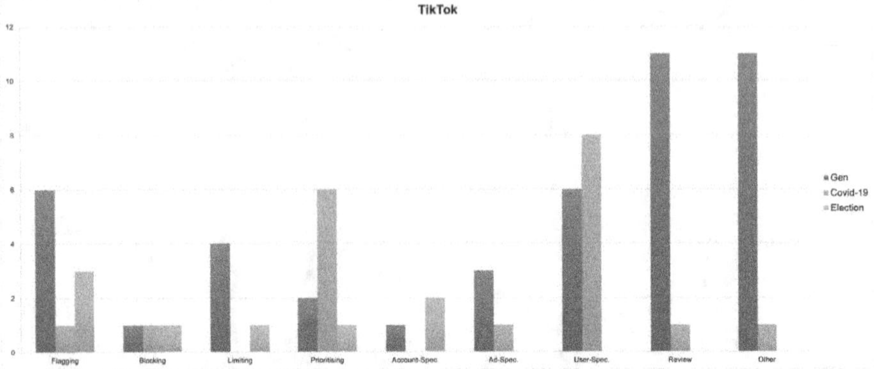


## Discussion

Despite the lack of a unified response to online misinformation (as illustrated above), the online platforms examined in this study share some common aspects. We make four observations that demonstrate tendencies and enduring challenges in (audited self-) regulation of social media platforms and make initial comparisons against the renewed Code of Practice on Disinformation and the proposed Regulation on Political Advertising.

First, misinformation with "an imminent risk of physical harm and seeking to malignly influence political processes" is commonly considered content that needs to be deleted from the platform, whereas misinformation that falls outside of this category is generally allowed (Google, n.d.-e; Meta, n.d.-a; TikTok, n.d.; Twitter, n.d.-b). Harm, influence and intent are hard to establish, however, and what constitutes harmful content differs across platforms.

Indeed, platforms' threshold for harmful content differs. When does content seek to wilfully harm or undermine public trust, when is this content protected by the right to freedom of opinion and expression, and are they even mutually exclusive? While recognising the difficulty, the political stakes and the criticism to which platforms expose themselves by being transparent about reactive measures, we can say that additional guidance and examples on clear boundaries for users, especially in electoral contexts, are needed. We positively note that the revised Code calls upon signatory platforms to commonly define misinformation, disinformation and set clear guidelines for what constitutes harmful misinformation (European Commission, 2022).[13] The closer involvement of fact-checkers and other civil society is also a positive development, both for transparency and consultation purposes. How robust and enforceable the Code is (especially through the regulatory teeth of the Digital Services Act for Very Large Online Platforms) will become clear in the coming months.

In the same vein, there is still a lot of leeway for the platforms to define what constitutes 'political' (Cipers and Meyer, 2022). The lack of a provided definition for 'political' advertisements allows the different companies to approach this very differently: while Google limits its curational measures to electoral ads, Meta and Twitter treat advertisements of issue-based non-profit organisations the same way as those run by political parties or interest groups.[14]

Second, platform measures are heavily influenced by ongoing events. Platform measures focus mostly on misinformation surrounding the COVID-19 pandemic and elections (more recently -although not included in this research- the Russian invasion in the Ukraine). Meta, Google, and Twitter also contextualise their misinformation policies according to current events (this seems less the case with TikTok). In our analysis, most platform measures related to elections and political campaigning were set up in the context of US elections, followed by some measures in preparation of French, EU and Brazilian elections.[15] We see that effort was made in preparation for the French presidential elections in 2022, however little in comparison to the U.S. presidential elections (Meta, n.d.-c). In our view, platforms' ability to react swiftly with new measures and updated policies is neither inherently positive nor negative. It is a feature that demonstrates their intermediary role in speech regulation.

---

13 In our view, an exception to these rather vague community guidelines and terms of service might be Twitter's clarification on conditions for taking action against accounts that share misinformation that could bring harm to crisis-affected populations. For conflict-related content to be in violation of Twitter's crisis misinformation policy, it must: *"advance a claim of fact, expressed in definitive terms; be demonstrably false or misleading, based on widely available, authoritative sources; and be likely to impact public safety or cause serious harm"* (Twitter, n.d.-d). This definition holds into account intent ("a claim of fact, expressed in definitive terms"), what constitutes misinformation ("demonstrably false or misleading, based on widely available, authoritative sources") and possible outcomes ("likely to impact public safety or cause serious harm").

14 For more information on the differences between social media companies' political advertisement policies, see Cipers and Meyer (2022).

15 It should be noted that neither our focus as researchers (in this article) nor the focus of the platforms (in their reporting to the European Commission) were global in scope. We acknowledge that platforms seek to help with electoral processes in democracies globally, as for instance reported in Brown et al. (2020).

Third, as clearly demonstrated in the analysis of this article, online platforms emphasise preventative measures. For all studied platforms, we noticed that measures connecting users to authoritative sources, helping them navigate a crisis or decision (elections or vaccines) and other measures highlighting technical advantages and solutions to outlined problems (preventative measures) were more prevalent and explained more in depth in the reports and in the communication to users. Measures that block content and users or measures that limit the reach of content or users are less pronounced in platform reporting. Of course, platforms will not be eager to admit or appear to be solving problems that they helped produce due to a lack of intervention in the past. Instead, they profile themselves as helpful tools and companies that fare better under self/co-regulation rather than having to perform reactive content moderation under pressure from political institutions. Paradoxically, their extensive use and promotion of preventative measures send the signal that platforms do not wish to impede the liberal rights granted to their users, while confirming that they are not neutral platforms, but proactively moderate content and shape discourses.

Fourth, rules for political advertising differ profoundly. Twitter and TikTok opt to ban political advertisements from their platforms all together (Dang, 2020; Twitter, n.d.-c), while companies like Meta seek middle ground and provide labels to alert users when they see a political advertisement (Meta, n.d.-d). We applaud the EU's endeavour to regulate political advertising but also question its limited focus in a changing social media landscape: pages, groups and other intra-platform networks can influence voter behaviour and political disposition without the need of paid advertisements or boosted content. At the same time, we recognise that if platforms would unilaterally decide to moderate (political) misinformation more harshly, this could trigger cross-migration of users who infringe terms of service and move to another platform with less strict policies, thereby generating a whack-a-mole situation in dealing with harmful content and behaviour – and potentially feeding polarisation of political discourses (Rogers, 2020). Content moderation in itself is a political issue (Alizadeh et al., 2022). Uncoordinated ad hoc platform measures provide ample room for abusive and manipulative use and manoeuvring by politicians and political parties. The proposed regulation on political advertising is a big step forward as it limits the paid reach of political speech, but in our view, it is the first step of a long journey.

## Conclusion

Circling back to this article's main research question on how platforms self-regulate online political campaigning through their policies and practices related to (political) misinformation and advertising, our finding confirms that ad-specific policies play an important role in platforms' moderation.

Meta and Google are particularly active in using their ad policies to regulate online political campaigning. Their political/electoral ad-specific measures focus on creating transparency through Ad Libraries; they do not address the targeting of political subjects at this time.[16] Although Twitter and TikTok take far less ad-specific measures related to political campaigning, their measures are stronger, by simply banning political advertising altogether. In addition, platforms label political and government accounts, which is similarly to the Ad Libraries, a transparency measure that does not touch polarising political speech as such. Indeed, in the case of Meta, this even creates an exception status, shielding politicians from third party fact-checking.

While cumulatively preventative measures to counter misinformation prevail for platforms, platforms are less hesitant to use reactive measures, including blocking or limiting content, to handle political/electoral misinformation, than we initially expected. However, as the focus is limited to content that can cause significant harm to others and undermine trust in civic institutions, many shades of polarising content are left untouched. Indeed, Meta and Twitter allow exceptions for content that potentially goes against their policies based on newsworthiness and perceived public interest.

Thus, returning to two core arguments: platform policies and practices merit analytical scrutiny as they determine the boundaries of acceptable speech online; and measures to regulate political speech should be understood as going beyond political advertising alone, but also include policies regarding political accounts and political/election misinformation. We find that platforms are still in a position where they can establish their own normative interpretation of what constitutes misinformation, when it is (un)acceptable and how it should be tackled. They draw the boundaries of acceptable speech

16 Interestingly, ad-specific measures were also extensively used during the pandemic. Public health ad-specific measures mostly pertained to providing free advertising credits to public health professionals and governments in order to ensure access to reliable and authoritative content – thus overlapping with prioritising/amplifying content, but as the measures were ad-specific, they were categorised separately (Meyer and Alaphilippe, 2021).

and behaviour, even after the revisions to the Code of Practice on Disinformation and during the ongoing discussions on the Regulation on the Transparency and Targeting of Political Advertising. Political "free speech may not be free reach" (DiResta, 2018) for users on some platforms, but the free reign of platforms still largely remains.

## Bibliography

### Scientific literature

**Alizadeh, Meysam, Gilardi, Fabrizio, Hoes, Emma, Klüser, K. Jonathan, Kubli, Maël and Marchal, Nahema** (2022), "Content moderation as a political issue: the Twitter discourse around Trump's ban", *Journal of Quantitative Description: Digital Media*, vol. 2, p. 1-44, <DOI: 10.51685/jqd.2022.023>.

**Davis, Mark** (2021), "The online anti-public sphere", *European Journal of Cultural Studies*, vol. 24 (1), p. 143-159, <DOI: 10.1177/1367549420902799>.

**Gillespie, Tarleton** (2018), *Custodians of the internet. Platforms, content moderation, and the hidden decisions that shape social media*, New Haven, Yale University Press.

**Gorwa, Robert** (2019), "What is platform governance?", *Information, Communication & Society*, vol. 22 (6), p. 854-871, <DOI:10.1080/1369118X.2019.1573914>.

**Gorwa, Robert, Binns, Reuben and Katzenbach, Christian** (2020), "Algorithmic content moderation: technical and political challenges in the automation of platform governance", *Big Data & Society*, vol. 7 (1), p. 1-15, <DOI: 10.1177/2053951719897945>.

**Haggart, Blayne and Iglesias Keller, Clara** (2021), "Democratic legitimacy in global platform governance", *Telecommunications Policy*, vol. 45 (6), p. 102-152, <DOI: 10.1016/j.telpol.2021.102152>.

**Häussler, Thomas** (2017), *The media and the public sphere: a deliberative model of democracy*, London, Routledge.

**Helberger, Natali, Dobber, Tom and de Vreese, Claes** (2021), "Towards unfair political practices law: learning lessons from the regulation of unfair commercial practices for online political advertising", *Journal of Intellectual Property, Information Technology and Electronic Commerce Law*, vol. 12 (3), p. 273-296.

**Kaye, David** (2019), *Speech police: the global struggle to govern the internet*, New York, Columbia Global Reports.

**Kettemann, Matthias C.** (2020), *The normative order of the internet: a theory of rule and regulation online*, Oxford, Oxford University Press.

**Marsden, Chris, Meyer, Trisha and Brown, Ian** (2020), "Platform values and democratic elections: How can the law regulate digital disinformation?", *Computer Law & Security Review*, vol. 36, 105373, <DOI: 10.1016/j.clsr.2019.105373>.

**Meyer, Trisha and Alaphilippe, Alexandre** (2021), "Platform (un)accountability. Reviewing platform responses to the global disinfodemic one year onward", *in* Judit Bayer, Bernd Holznagel, Päivi Korpisaari and Lorna Woods (eds), *Perspectives on platform regulation. Concepts and models of social media governance across the globe*, Baden-Baden, Nomos, p. 509-530.

**O'Regan, Catherine** (2018), "Hate speech online: an (intractable) contemporary challenge?", *Current Legal Problems*, vol. 71 (1), p. 403-429, <DOI: 10.1093/clp/cuy012>.

**Papacharissi, Zizi** (2002), "The virtual sphere: the internet as a public sphere", *New Media and Society*, vol. 4 (1), p. 9-27, <DOI: 10.1177/14614440222226244>.

**Rogers, Richard** (2020), "Deplatforming: Following extreme Internet celebrities to Telegram and alternative social media", *European Journal of Communication*, vol. 35 (3), p. 213-229, <DOI: 10.1177/0267323120922066>.

**Stockmann, Daniela** (2023), "Tech companies and the public interest: the role of the state in governing social media platforms", *Information, Communication & Society*, vol. 26 (1), p. 1-15.

**Zuboff, Shoshana** (2019), *The age of surveillance capitalism: The fight for a human future at the new frontier of power*, New York, Public Affairs.

## Grey literature

**Bontcheva, Kalina, Posetti, Julie, Teyssou, Denis, Meyer, Trisha, Gregory, Sam, Hanot, Clara and Maynard, Diana** (2020), *Balancing act: Countering digital disinformation while respecting freedom of expression*, UNESCO, <https://en.unesco.org/publications/balanceact>.

**Brown, Ian, Marsden, Christopher, Lee, James and Veale, Michael** (2020), *Cybersecurity for elections. A Commonwealth guide on best practice*, London, Commonwealth Secretariat, <https://thecommonwealth.org/news/commonwealth-launches-guide-tackle-election-cybersecurity-threat>.

**Cipers, Samuel and Meyer, Trisha** (2022), "What is political? The uncoordinated efforts of social media platforms on political advertising", *EDMO BELUX*, 19 December, <https://belux.edmo.eu/what-is-political-the-uncoordinated-efforts-of-social-media-platforms-on-political-advertising>.

**Culliford, Elizabeth** (2022), "Facebook-owner Meta to share more political ad targeting data", *Reuters*, 23 May, <https://www.reuters.com/technology/facebook-owner-meta-share-more-political-ad-targeting-data-2022-05-23/>.

**Dang, Sheila** (2022), "TikTok to clamp down on paid political posts by influencers ahead of U.S. midterms", *Reuters*, 17 August, <https://www.reuters.com/world/us/tiktok-clamp-down-paid-political-posts-by-influencers-ahead-us-midterms-2022-08-17/>.

**DiResta, Renee** (2018), "Free Speech Is Not the Same As Free Reach", *WIRED*, 30 August, <https://www.wired.com/story/free-speech-is-not-the-same-as-free-reach/>.

**European Commission** (2018a), "Communication from the Commission to the European Parliament, the Council, the European Economic and Social Committee and the Committee of the Regions. Tackling online disinformation: a European Approach", COM(2018) 236 final, Brussels, 26 April, <https://eur-lex.europa.eu/legal-content/EN/TXT/?uri=CELEX:52018DC0236>.

**European Commission** (2018b), *EU Code of Practice on Disinformation*, European Commission, <https://digital-strategy.ec.europa.eu/en/library/2018-code-practice-disinformation>.

**European Commission** (2020a), "Communication from the Commission to the European Parliament, the Council, the European Economic and Social Committee and the Committee of the Regions on the European democracy action plan", COM(2020)790 final, Brussels, 3 December, <https://ec.europa.eu/info/strategy/priorities-2019-2024/new-push-european-democracy/european-democracy-action-plan_en>.

**European Commission** (2020b), *European Democracy Action Plan: making EU democracies stronger* [Press release], IP 20 2250, Brussels, 3 December, <https://ec.europa.eu/commission/presscorner/detail/en/ip_20_2250>.

**European Commission** (2021), "Proposal for a Regulation of the European Parliament and the Council on the transparency and targeting of political advertising", COM(2021)731 final, Brussels, 25 November, <https://ec.europa.eu/info/policies/justice-and-fundamental-rights/eu-citizenship/democracy-and-electoral-rights_en>.

**European Commission** (2022), *2022 EU Code of Practice on Disinformation*, European Commission, <https://digital-strategy.ec.europa.eu/en/policies/code-practice-disinformation>.

**Google** (n.d.-a), *COVID-19 medical misinformation policy*, YouTube help, <https://support.google.com/youtube/answer/9891785?hl=en>.

**Google** (n.d.-b), *Political ads on Google*, Ads transparency Google, <https://adstransparency.google.com/political?political=&region=BE>.

**Google** (n.d.-c), *Political content*, Advertising policies help, <https://support.google.com/adspolicy/answer/6014595#zippy=%2Ceuropean-union-eu-election-ads%2Caustralia-election-ads%2Cunited-states-us-election-ads>.

**Google** (n.d.-d), *Elections misinformation policies*, YouTube help, <https://support.google.com/youtube/answer/10835034?hl=en&ref_topic=10833358#zippy=%2Cvoter-suppression%2Cincitement-to-interfere-with-democratic-processes>.

**Google** (n.d.-e), *Misinformation policies*, YouTube help, <https://support.google.com/youtube/answer/10834785?hl=en&ref_topic=10833358>.

**Harrison, Mantas** (2021), "Facebook acknowledges politicians can harm, but won't let them be fact-checked", *Poytner*, 10 June, <https://www.poynter.org/fact-checking/2021/facebook-acknowledges-politicians-can-harm-but-wont-let-them-be-fact-checked/>.

**Hate Aid** (2021), *Boundless hate on the internet – Dramatic situation across Europe*, Berlin, 3 November, <https://hateaid.org/wp-content/uploads/2022/04/HateAid-Report-2021_EN.pdf>.

**Meta** (2022), *Our approach to newsworthy content*, Meta transparency center, 25 August, <https://transparency.fb.com/en-gb/features/approach-to-newsworthy-content/>.

**Meta** (n.d.-a), *Community Standard*, Facebook community standards, <https://transparency.fb.com/nl-nl/policies/community-standards/misinformation/>.

**Meta** (n.d.-b), *We're committed to supporting COVID-19 relief efforts*, <https://about.facebook.com/actions/responding-to-covid-19/>.

**Meta** (n.d.-c), *Preparing for elections*, <https://about.facebook.com/actions/preparing-for-elections-on-facebook/>.

**Meta** (n.d.-d), *How disclaimers work for ads about social issues, elections or politics*, Meta business help center, <https://www.facebook.com/business/help/198009284345835?id=288762101909005>.

**Meyer, Trisha and Pershan, Claire** (2022), "Room for improvement. Analysing redress policy on Facebook, Instagram, YouTube and Twitter", *EU Disinfolab*, 7 June, <https://www.disinfo.eu/publications/room-for-improvement-analysing-redress-policy-on-facebook-instagram-youtube-and-twitter/>.

**Posetti, Julie, Aboulez, Nermine, Bontcheva, Kalina, Harrison, Jackie and Waisbord, Silvio** (2020), "Online violence against women journalists: a global snapshot of incidence and impacts", CI/FEJ/2020/PI/1, UNESCO, <https://unesdoc.unesco.org/ark:/48223/pf0000375136>.

**Posetti, Julie and Bontcheva, Kalina** (2020), "Disinfodemic: deciphering COVID-19 disinformation", *Policy brief 1*, UNESCO, <https://en.unesco.org/covid19/disinfodemic/brief1>.

**TikTok** (n.d.), *Community guidelines*, TikTok community guidelines, <https://www.tiktok.com/community-guidelines?lang=en#29>.

**TikTok** (2022), *Our commitment to election integrity*, TikTok newsroom, 17 August, <https://newsroom.tiktok.com/en-us/our-commitment-to-election-integrity>.

**Twitter** (2021), *COVID-19 misleading information policy*, Twitter help center, December, <https://help.twitter.com/en/rules-and-policies/medical-misinformation-policy>.

**Twitter** (n.d.-a), *Coordinated harmful activity*, Twitter Help Center, <https://help.twitter.com/en/rules-and-policies/coordinated-harmful-activity>.

**Twitter** (n.d.-b), *Civic integrity policy*, Twitter help center, <https://help.twitter.com/en/rules-and-policies/election-integrity-policy>.

**Twitter** (n.d.-c), *Political content*, Twitter business, <https://business.twitter.com/en/help/ads-policies/ads-content-policies/political-content.html>.

**Twitter.** (n.d.-d), *Crisis misinformation policy*, Twitter help center, <https://help.twitter.com/en/rules-and-policies/crisis-misinformation>.

**Tworek, Heidi** (2019), "Social media platforms and the upside of ignorance", *Centre for International Governance Innovation*, 9 September, <https://www.cigionline.org/articles/social-media-platforms-and-upside-ignorance>.

**UN Special Rapporteur on the Promotion and Protection of the Right to Freedom of Opinion and Expression** (2018), *A/HRC/38/35. Report on content regulation*, United Nations, Office of the High Commissioner for Human Rights, <https://www.ohchr.org/en/calls-for-input/report-content-regulation>.

**UN Special Rapporteur on the Promotion and Protection of the Right to Freedom of Opinion and Expression** (2019), *A/74/486. Report on online hate speech*, United Nations, Office of the High Commissioner for Human Rights, <https://unesdoc.unesco.org/ark:/48223/pf0000375136>.

**Wardle, Claire and Derakhshan, Hossein** (2017), *Information Disorder: Toward an interdisciplinary framework for research and policy making*, DGI(2017)09, Strasbourg, Council of Europe, <http://tverezo.info/wp-content/uploads/2017/11/PREMS-162317-GBR-2018-Report-desinformation-A4-BAT.pdf>.

**Samuel Cipers** is a PhD researcher at Brussels School of Governance's Centre for Digitalisation, Democracy and Innovation (Vrije Universiteit Brussel) and the Department of Political Science at the University of Warwick. His research focusses on establishing a critical framework to study the European Union's response to online disinformation and populist discourse on- and offline by using poststructuralist discourse analysis and other critical frameworks.

samuel.cipers@vub.be

**Trisha Meyer** is an Assistant Professor in Digital Governance and Participation at the Vrije Universiteit Brussel and a Professorial Fellow at the United Nations University CRIS in Bruges. Trisha leads the Research Centre on Digitalisation, Democracy and Innovation at VUB Brussels School of Governance and is the principal investigator of EDMO BELUX, an EU-funded hub on research, fact-checking and media literacy on online disinformation in Belgium and Luxembourg. Her research focuses on platform moderation policies and practices as they intersect with illegal and harmful content.

trisha.meyer@vub.be

## Annex A: Overview of longitudinal monitoring of platform responses to fight misinformation (separate file)

## Annex B: Overview of consulted platform community guidelines, terms of service and blogs (list)

### Meta (Facebook & Instagram)

A. Blogs & Reports

- Blog on commitment to support COVID-19 relief efforts <https://about.facebook.com/actions/responding-to-covid-19/>
- Blog on actions in the face of elections <https://about.facebook.com/actions/preparing-for-elections-on-facebook/>

B. Community Standards & Terms of Service

- Facebook's community standards <https://transparency.fb.com/nl-nl/policies/community-standards/?source=https%3A%2F%2Fwww.facebook.com%2Fcommunitystandards%2Fintroduction>
- Facebook's community standards enforcement report <https://transparency.fb.com/data/community-standards-enforcement/?source=https%3A%2F%2Ftransparency.facebook.com%2Fcommunity-standards-enforcement>
- Instagram's community guidelines <https://help.instagram.com/477434105621119/?helpref=hc_fnav>
- Instagram's terms of use <https://help.instagram.com/477434105621119/?helpref=hc_fnav>

### Google (Google & YouTube)

A. Blogs & Reports

- Google's blog <https://about.google>
- Google's crisis response blog <https://crisisresponse.google/?utm_source=about&utm_medium=referral&utm_campaign=footer-link>

B. Community Standards & Terms of Service Reports

- Google's transparency report <https://transparencyreport.google.com>
- YouTube's community guidelines <https://www.youtube.com/howyou-

tubeworks/policies/community-guidelines/#community-guidelines>

- YouTube's policies <https://support.google.com/youtube/topic/2803176?hl=en&ref_topic=6151248>
- Google's political ads transparency report <https://adstransparency.google.com/political?political=&region=BE>
- Google's ad policies for political content <https://support.google.com/adspolicy/answer/6014595#zippy=%2Ceuropean-union-eu-election-ads%2Caustralia-election-ads%2Cunited-states-us-election-ads>
- YouTube's elections misinformation policies <https://support.google.com/youtube/answer/10835034?hl=en>
- Google's privacy and terms of service <https://policies.google.com/pri-vacy>
- Google's community standards <https://about.google/community-guide-lines/>

## Twitter

A. Blogs & Reports

- Twitter's blog on COVID-19 <https://blog.twitter.com/en_us/topics/company/2020/covid-19>

B. Community Standards & Terms of Service

- Twitter's help center (rules & policies) <https://help.twitter.com/en/rules-and-policies>
- Twitter's transparency reports <https://transparency.twitter.com/en.html>
- Twitter's policies on political ads <https://business.twitter.com/en/help/ads-policies/ads-content-policies/political-content.html>

## TikTok

A. Blogs & Reports

- TikTok's blog (newsroom) <https://newsroom.tiktok.com/en-us>

B. Community Standards & Terms of Service

- TikTok's community guidelines <https://www.tiktok.com/community-guidelines?lang=en#37>

POLITIQUE EUROPÉENNE
N° 81 | 2023
Sebastian Heidebrecht
p. 142-168

**Platform Accountability in the European Union**

**The Cases of Data Protection and Digital Services Regulation**

Big tech companies mimic competencies that have been the prerogative of nation-states. Many high-profile public scandals fuelled calls for more accountability. Although the EU has addressed the topic, we do not know yet how the accountability of platform companies increased. This article develops the concept of *platform accountability* as an analytical tool. An analysis of three influential regulations shows that data protection regulation increases democratic accountability but has problems with effective enforcement. Recent digital services regulations are mending this weakness and ultimately empowering the Commission. Yet, the Commission has to prove that it can act as a credible digital regulator.

**Responsabilité des plateformes dans l'Union européenne**

**Les cas de protection des données et de réglementation des services numériques**

Les grandes entreprises technologiques imitent des compétences qui étaient la prérogative des États-nations. De nombreux scandales publics très médiatisés ont alimenté les appels à davantage de responsabilité. Bien que l'UE ait abordé le sujet, nous ne savons pas encore comment la responsabilité des entreprises de plateforme a augmenté. Cet article développe le concept de responsabilité de plateforme en tant qu'outil analytique. Une analyse de trois réglementations influentes montre que la réglementation sur la protection des données accroît la responsabilité démocratique mais rencontre des problèmes d'application efficace. Les récentes réglementations sur les services numériques comblent cette faiblesse et, à terme, donnent du pouvoir à la Commission. Pourtant, la Commission doit prouver qu'elle peut agir en tant que régulateur numérique crédible.

# Platform Accountability in the European Union

## The Cases of Data Protection and Digital Services Regulation

**Sebastian Heidebrecht**
Centre for European Integration Research (EIF), Institute of Political Science, University of Vienna

## Introduction

Digital services are dominated by a few very large, predominantly North-American, platform companies. Enterprises like Meta (Facebook), Apple, Alphabet (Google), Microsoft, and Amazon benefit from economies of scale, high switching, and low marginal costs, which allows them not only to become the world's most valuable companies but also to transform into all-encompassing service providers, offering clients all sorts of products that range from finance to insurance to health care. Originally, platform business models and respective developments in the digital economy promoted overcoming European Union (EU) market fragmentation in the digital sphere. With the commercialization of the internet in the 1990s came the rise of these big technology companies. Customers across the single market, previously fragmented along national borders, could for example buy products on Amazon's marketplace, use the services linked to Google's search engine, and exchange and connect through Facebook's social media.

But, as it turns out, the market power of these companies has also a downside. Public scandals like the revelations of Edward Snowden in 2013 and Frances Haugen in 2021 appear only as the tip of the iceberg of many problematic practices of large platform companies and their managers. Many customers and policymakers alike became aware of the significant powers of, and risk of misconduct by big technology companies. This added to a shift in the public discourse about big technology companies which no longer focuses on the liberalising potential of free internet, but points to the negative effects that are associated with large platform business models, thereby evoking a "darker narrative of platform capitalism" (Pasquale, 2016, p. 314). In parallel, demands for more platform accountability have become louder over the last few years (e.g. Helberger, Pierson, and Poell, 2018; Lehdonvirta, 2022; Suzor, 2019). In the EU, policymakers aim to ensure a safe and accountable online environment (European Commission, 2022) and to rebalance the responsibilities of users, platforms, and public authorities according to European values (European Commission, 2022). Furthermore, even before the Russian invasion to Ukraine, the EU has aimed to increase control over the digital sphere and to build digital sovereignty (Floridi, 2020; Obendiek, 2021).

This article is about the EU's attempt to hold large platform companies accountable. This issue is important because of the big technology companies' size, power and business models. In many ways, large online platforms have emerged as dominant private monopolies, more akin to the provision of public infrastructure than competitive markets (Rahman, 2017; van Dijck, Nieborg, and Poell, 2019). Customers often no longer have a fair opportunity to choose another option. Furthermore, their concentrated economic power leads to significant political power over entire economic sectors and public opinion (Helberger, 2020), so large online platforms are portrayed as overtaking the state (Lehdonvirta, 2022) and deemed to be governors of online speech (Klonick, 2018). The threats that emanate from Big Tech using big data are not just market dominance but also the power to give advantages to one group, such as sellers, over others, such as buyers. When the problem is not only market dominance but power, the response should be found in accountability, not just competition policy (Pistor, 2020).

In this article, I demonstrate the added value of adjusting an acknowledged assessment tool developed by Mark Bovens, Thomas Schillemans and Paul t'Hart which – broadly speaking – identifies evaluative questions that can be used to assess platform accountability (e.g. Bovens, Schillemans, and t'Hart, 2008; Schmidt, 2013; Schmitter, 2004). On a conceptual level, the article contributes to the more general debate about understanding platform governance (Gorwa, 2019). It develops the concept of platform accountability that is distinct from other but related concepts such as platform's transparency and legitimacy. For example, transparency mostly refers to manifold ways of accessing information and legitimacy refers to a more normative assessment of platform decisions. By contrast, platform accountability refers to a specific set of institutionalised practices, mostly in the form of laws, to hold platforms to account. There are different proposals for which set of actors should hold platforms accountable, like users, stakeholders or citizens more generally (Haggart and Keller, 2021; Pistor, 2020; van Dijck et al., 2019). My concept of platform accountability centres on a specific understanding of public accountability and thus on the legal arrangements that are implemented in the EU.

By assessing platform accountability in three influential EU regulations (the General Data Protection Regulation, the Digital Services Act and the Digital Markets Act), my article contributes to the emerging debate about developments in EU digital policy more generally (e.g. Falkner, Heidebrecht, Obendiek, and Seidl, 2022; Heidebrecht, 2023; Obendiek, 2021). I distinguish between more democratic and more constitutional features of platform accountability and show that the EU's most recent attempts at digital services regulations are informed by previous weaknesses in the constitutional dimension of platform accountability, notably in the area of data protection. However, the three regulations together have the potential to form a coherent and more robust framework for platform accountability. Because this rests ultimately on an empowered European Commission, I also outline reasons why the Commission has yet to prove that it can act as a credible regulator to hold platform companies to account.

The question of how the EU holds large online platforms to account is of immense importance for democracies, economies, and societies in the 21st century. Given the EU's attempt to be both a market and a normative power in world politics (Damro, 2012; Manners, 2002) and in digital regulation alike (Bradford, 2020), knowing how it is designing platform accountability is important for understanding the politics of shaping the global digital sphere. Before I present my findings on the EU's attempt to ensure platform accountability in three case studies, I provide a discussion of both the need for platform accountability and the EU's history of creating it in the light of the existing literature, detail my theoretical approach and elaborate on my empirical strategy. The last section concludes.

## Platform Power and the Need for Accountability

To address the issue of platform accountability, I briefly discuss what platforms are and why accountability is necessary. Platform companies are companies that operate a variety of platforms, ranging from social media platforms like Facebook to video platforms like YouTube, to other services or marketplaces and search engines, which are operated typically by big companies. Some of these platforms are entangled in virtually every aspect of contemporary life, from politics and industrial relations to cultural production and consumption (Gillespie, 2018; Nieborg and Poell, 2018; van Doorn, 2017). Although many of the services platforms offer have been praised as 'liberalisation technologies' (Tucker, Theocharis, Roberts, and Barberá 2017), the sector is dominated by a handful of large, predominantly American technology firms. Because of the dominance of this handful of American companies, many concerns have been raised in scholarly discourse, not least in the aftermath of many high-profile public scandals (Vaidhyanathan, 2018), such as the revelations by Edward Snowden in 2013 (Mazzetti and Schmidt, 2013), highlighting the problematic practices of many big US technology companies and intelligence services.

Against this backdrop, scholarly discourse addressed many of the negative consequences that are associated with an unregulated digital economy. John W. Cioffi, Martin F. Kenney and John Zysman (2022) point out that the minimalist regulatory regime led to economic, political and social challenges. In the economic dimension, scholars emphasize that platform companies constrain competitive markets through multiple forms of anti-competitive behaviour (Khan, 2017), so that some have argued for breaking up these companies or preventing them from making future acquisitions (Pasquale, 2018; Wu, 2018). The involvement of the private US company Cambridge Analytica in the 2016 US election campaign (Chen, 2018), and Russian interference in the 2016 US election (Abrams, 2019) shone a light on political challenges, that were seen as examples of potential political threats to the European elections in 2019 (Plucinska, 2018). Because today many people derive their political information from online sources, platform companies in many respects constitute the infrastructure of public discourse and free speech (Klonick, 2018). As private companies that are profit-driven rather than focussed on ensuring democracy, their business models endanger democratic discourse through the spread of disinformation and hate speech (Howard, 2020) and harm the privacy and fundamental rights of citizens (Zuboff, 2019).

Given their monopolistic position, very large online platforms are frequently compared to public infrastructure for economic, political and social life (van Dijck et al., 2019) and the private companies that create and control that infrastructure are described in terms of the "new governors in the digital era" (Klonick, 2018, p. 1663). This leads to a form of "opinion power" (Helberger, 2020, p. 842) or, in the words of Manuel Castells, "the capacity to influence people's minds" (cited in Helberger, 2020, p. 845). They can do so by controlling access to goods that comprise a backbone of much of modern social and economic activity, upon which many communities and constituencies depend

(Rahman, 2017, p. 1622). Large online platforms possess important qualities because they control the digital ecosystem. This amounts to significant infrastructural power (Valdez, 2023; van Dijck et al., 2019).

Against this backdrop, it is noteworthy that the choice between alternative institutional frameworks is not an individual choice that can be resolved on a market, but a collective choice, which is precisely why political institutions matter in this debate. It is important to understand that in exercising their powers, very large online platforms frequently evoke competences that were previously considered to be the prerogative of the nation state. One reason is their size. Amazon, for example, had an estimated USD 490 billion's worth of goods passing through its platform, more than many countries' gross domestic product. The company earned almost USD 75 billion in fees from merchants who used its marketplace and logistics infrastructure – more than most governments collect in taxes (Lehdonvirta, 2022, p. 3). Another example of how large platform companies are emulating the nation state is the issuing and enforcing of laws. They define the terms and conditions of economic exchange on their platforms as well as settle conflict between opposing parties. For example, eBay (an US company) alone claimed that it resolved more than 60 million disputes in a single year, which is not very far away from the 90 million cases in US courts (Lehdonvirta, 2022, p. 2).

According to liberal reasoning as well as democratic theory, large power requires checks and balances. In the case of large online platforms, they have to be found in the area of accountability because of the digital market structure, large platform companies' business models, and the infrastructural quality of their services. Checks and balances of corporate digital power will not come from competition, laissez-faire regulation, and market-based equilibria as the emergence of hegemonic positions and de facto monopolies demonstrates (Floridi, 2020, p. 372). Furthermore, the business models of large online platforms do not just rely on market dominance but on the power to bestow advantages upon one group, such as sellers, at the expense of other groups, such as buyers (Pistor, 2020). Thus, we leave the area of markets and enter the area of power, which is why we need the toolkit of accountability rather than for example competition law. Of course, platform companies are already in some respect responsive to the demands of the customers, but they are so, as Blayne Haggart and Clara Iglesias Keller (2021, p. 6) put it nicely, "in the same way that a monarch [is] responding to the pleas of his subjects, [which should ...] not be confused with a form of democracy".

## How the European Union Holds Platforms to Account

In the early days of the commercial Internet, EU policymakers adopted a pronounced market liberal regulative approach to the emerging digital sphere. Private technology companies on liberal markets were considered as promoting a "market-driven revolution" (European Council, 1994) to achieve European competitiveness in a global information society. Of course, the EU was from early on concerned with questions of data protection as the Data Protection Directive from 1995 demonstrates (Directive 95/46/EC, see also Newman, 2008). Other areas such as digital services legislation in form of the 2000 e-commerce directive (Directive 2000/31/EC) ensured light touch regulation (e.g. Farrand, 2023) and established important principles like limited liability of digital service providers for online content (Heidebrecht, 2023). While the EU took in some respects a dual-approach that combines aspects such as consumer protection with policies that were geared to promote market activity (Newman, 2020), public accountability of large technology companies and platform operators was very limited.

Approximately over the last decade, the EU developed a relatively comprehensive update of key legislations that affect the digital sphere. The Juncker Commission promised with its 2015 Digital Single Market Strategy (European Commission, 2015) to assess the role of large online platforms. The "subsequent 2016 report" (European Commission, 2016) on the issue identified in particular challenges in the areas of transparency, competition and content moderation. The 1995 Data Protection Directive was updated with the 2016 General Data Protection Regulation (Regulation (EU) 2016/679, or GDPR), frequently referred to as the global gold standard (Schünemann and Windwehr, 2021). The issue of transparency of large online platforms was addressed by a 2019 Platform to Business Regulation (Regulation (EU) 2019/1150), and the 2022 digital services package deals with the issues of competition and content moderation. The package was welcomed by the Commission's Executive Vice-President, Margrethe Vestager, for its "help [to] create a safe and accountable online environment [...] and ensure that platforms are held accountable for the risks their services can pose to society and citizens" (Vestager, 2022). The responsible minister of the Czech Council Presidency, Ivan Bartoš, in charge of the agreement on the Digital Markets Act, argued that it "will finally make large online platforms responsible for their actions" (Bartoš, 2022). Public media covered the digital services package in terms of "a constitution of the internet" (Bertuzzi, 2021).

In academic discourse, the provisions of the Digital Services Package were seen as a comprehensive move towards encompassing socio-economic regulation and market intervention (Cioffi et al., 2022). This legislative package further moves digital competition policy towards a stronger approach that does no longer handle ex-post competition infringements but includes ex-ante provisions that also address market structure, thus implying provisions of broader economic regulation to ensure fair competition online (Cini and Czulno, 2022; Heidebrecht, 2022; Meunier and Mickus, 2020). In this regard, scholars noted a move away from the EU's alleged "neoliberal bias" (Laurer and Seidl, 2021, p. 257), in parallel with the more geopolitical rhetoric as it is expressed in the frequent usage of concepts such as open strategic autonomy and digital sovereignty (Falkner et al., 2022; Heidebrecht, 2023; Schmitz and Seidl, 2022). Against this backdrop, I develop the assessment tools for a potentially strengthened public form of platform accountability in the next section.

## Defining Platform Accountability

In the political science literature, there is an established debate about accountability which can help us understand the importance of the concept and its distinctness from others. It is important to note that accountability concerns those who exercise power, not those who are subordinate to it (Schedler, 1999, p. 20). Andreas Schedler (1999) describes accountability as a two-dimensional concept that implies answerability and enforcement. For him, there are three different ways of redressing the abuse of political power: threats of sanctions (enforcement), transparency (answerability) and justification (answerability). Thus, to be effective and perceived as such, accountability has to demonstrate that it can achieve its aim of curbing power. In this regard, accountability agencies must have not only legal authority but also sufficient and de facto autonomy to pursue their tasks (O'Donnell, 1998, p. 119). Effective accountability is considered important because it ensures participation of citizens and responsiveness of rulers. Failed accountability is considered leading to the unresponsive imposition of rules and the resentment of citizens (Schmitter, 2004).

In the debate about platform governance, regulation and accountability, Robert Gorwa (2019) distinguishes between three different approaches: platform self-governance, external governance exercised by public supervisors and the like, and platform co-governance by empowering users to participate in platform decision-making. In this terminology, the EU's early approach relied mostly on platform self-governance (Farrand, 2023) and a "laissez-faire" (Gorwa, 2019, p. 862) relationship between supervisors and platform companies. By contrast, more recent policy initiatives of the Commission aim to "rebalance [t]he responsibilities of users, platforms, and public authorities [...] according to European values, placing citizens at the centre" (European Commission, 2022). This aim indicates, using Gorwa's (2019) terminology, a shift from self- to external platform governance and has been described in terms of a change from a market-liberal towards a public-interventionist regulatory approach (Heidebrecht, 2023). It requires, in the words of the EU Commission, the establishment of "a common set of rules on intermediaries' obligations and accountability" (European Commission, 2022).

To assess the accountability mechanisms of external platform governance, this article develops the concept of a public form of *platform accountability* as an assessment tool. Haggart and Keller (2021) have demonstrated that the conceptual tools of the wider EU legitimacy and accountability literature (notably Scharpf, 1999; Schmidt, 2013) can be used to gain insights into the evolving debate about EU platform regulation. Different from their work, my empirical focus is on the legal mechanisms that are established to hold large online platforms to account and not on the democratic input or the results of the arrangements (basically what Schmidt 2013 and Haggart/Keller 2021 call throughput legitimacy).

Mark Bovens (2005, p. 184) defines accountability as "a social relationship in which an actor feels an obligation to explain [...] his or her conduct to some significant other". However, differences in the interpretation of accountability exist within different countries or regions. For example, in the Anglo-American discourse, accountability is often used as a conceptual umbrella that refers to more normative goods like transparency and trustworthiness. In this context, accountability has a more substantive dimension, that often refers to the quality of accountability arrangements assessed against the benchmark of normative goods. Bovens (2010, p. 948) calls such more evaluative analysis that asks if the actor's behaviour is in line with substantive standards "accountability as a virtue". By contrast, on the other side of the Atlantic, in British and continental European scholarly discourse, accountability is often used in more descriptive terms. In this context, accountability is presented in its procedural dimension as an institutional relation or arrangement in which an actor can be held to account by a forum (Bovens et al., 2008). A related and often more descriptive analysis commonly assesses the structure of the accountability arrangements. Bovens (2010, p. 950) calls this second perspective "accountability as a mechanism".

While accountability is related to legitimacy, it is also closely linked to, but should not be confused with, the concept of transparency. Transparency refers to principles of openness which are typically linked to and a prerequisite for responsiveness. This can be illustrated by the decision of Facebook to set up an Oversight Board on questions of content moderation in 2020. While the board can overrule decisions on single content moderation cases (Wong and Floridi, 2022), Facebook is not under a mandate to take up the board's broader recommendations. Thus, it creates transparency and leads to sort of an indirect accountability (Klonick, 2019) exercised through public pressure which requires Facebook to furnish reasons and explain its conduct. By contrast, more direct accountability by external governance bodies "demands some form of scrutiny by a specific forum" (Schmidt, 2013, p. 6). Different from transparency and platform self-governance, effective platform accountability would require demonstrating that decisions can be taken and enforced that are against the company's immediate best interests.

In the following, I approach big platform companies from a perspective that reflects their power, business model and kind of infrastructure-like services. Therefore, I apply a concept of accountability that covers the relationship of platforms to external actors and which is in the scholarly debate often referred to as "public accountability". Public accountability has two components. First, *public* accountability relates to openness. Explanations and justifications are given publicly, for example in the sense that they are open or at least accessible to citizens. The second element is *accountability.* As a concept, it is used in different areas, and in different settings like listed companies, in which shareholders implement mechanisms to hold the management to account, or in public policy, in which for example the executive branch of government is typically held accountable by parliaments as the representation of citizens (for an overview, see Bovens, Schillemans and Goodin, 2014). In this regard, my concept of a public form of platform accountability can be defined as follows: *Platform accountability refers to the legal mechanisms and institutional structures that shall ensure that public actors hold private platform companies to account.*

## Assessing Platform Accountability and Empirical Strategy

Because of the many powers large online platforms have in markets, societies and democracy, an analysis of platform accountability should cover both more substantive and more procedural dimensions. For this endeavour, I draw on the work of Bovens et al. (2008) who develop an assessment tool. They identify evaluative questions that can be used to assess if the accountability arrangements reflect different important qualities. In methodological terms, defining and using evaluative questions for different dimensions of public platform accountability will allow me to assess if EU regulations comply with these criteria and thus to describe the quality and processes of platform accountability in the EU.

To develop an assessment tool for public platform accountability, I adjust the assessment tool of Bovens et al. (2008). They distinguish between three perspectives on accountability of which I use the first two, namely: a democratic and a constitutional perspective (the third perspective requires analysing feedback-loops, and is less suitable for regulations that still need to be implemented). The perspective of democratic accountability is concerned with issues of popular control and the democratic chain of delegation and centres on democratically legitimized standards and preferences. To reflect the democratic accountability perspective, my evaluative question asks: Do accountability mechanisms accord with substantive standards and democratically legitimized preferences? The constitutional accountability perspective is primarily concerned with the issue of preventing power concentration and abuse and ensuring effective political and social control. To evaluate this constitutional perspective, I use two evaluative questions. First: Do accountability mechanisms allow monitoring and curtailing the abuse of powers and privilege? Second: Do public institutions possess credible enforcement mechanisms? See Table 1 for an overview.

**Table 1: Accountability Dimensions and Evaluative Questions**

| Dimension | Central Idea | Evaluative Questions |
|---|---|---|
| Democratic | *The accountability framework should reflect legitimised standards and preferences.* | *Do accountability mechanisms accord with substantive standards and democratically legitimized preferences?* |
| Constitutional | *The accountability framework should establish mechanisms to withstand power concentration and abuse and ensure effective political and social control.* | *Do accountability mechanisms allow monitoring and curtailing the abuse of powers and privilege?*<br>*Do public institutions possess credible enforcement mechanisms?* |

In order to assess platform accountability in the EU, I conduct a theory-guided analysis of three cases where I explore its democratic and constitutional dimensions.. In my case studies, I examine the legal characteristics of three important horizontal regulations that are set to shape how the EU regulates the digital sphere in important issue areas, namely data protection, digital services and digital markets. My study is primarily based on extensive document analysis of three EU regulations (see below). To assess platform accountability through these regulations, I used the three evaluative questions in a two-tier qualitative analysis. First, I checked the objectives of the regulations in their recitals (and compared this to selected press releases and media statements from the European Commission that accompanied the publication of the legislative proposals). Second, I used my evaluation questions as a guide for analysing the more substantive articles of the regulations, focusing on their more democratic and constitutional provisions that relate to the categories defined in my concept of platform accountability, such as enforcement articles and those codifying the more democratic or constitutional claims identified in the first step.

For these theory-guided qualitative case studies, I selected influential cases of EU legislation, which substantially affect platform companies in the EU's single market, by establishing horizontal and directly applicable rules across sectors of the digital economy. Choosing influential cases has typically the goal to explore their significance on some larger cross-case theory (Seawright and Gerring, 2008, p. 303). In my case, choosing influential cases allows me to demonstrate important patterns of regulatory reforms. While influential cases are typically not representative of a larger sample of EU digital regulation, my cases comprise the most far-reaching legislative initiatives in the area and thus allow me to draw conclusions on the development of EU platform accountability more generally.

In particular, I choose the most recent and most influential regulations in data protection, digital services and digital markets. The first case is the General Data Protection Regulation (GDPR), adopted in 2016. It creates strict data protection rules that are frequently referred to as the "gold standard of data protection" (Schünemann and Windwehr, 2021, p. 859). The second and third cases are the two regulations of the Digital Services Package, the Digital Services Act (DSA, Regulation (EU) 2022/2065) and the Digital Markets Act (DMA, Regulation (EU) 2022/1925). Substantive steps on both files between the EP and the Council were negotiated during the French Council Presidency from January to June 2022 and the digital services package was adopted in the first reading by the European Parliament in July 2022.

## Assessing Platform Accountability

### Platform Accountability in the General Data Protection Regulation

#### Democratic Public Accountability through the GDPR

The GDPR, adopted in 2016, builds up on the Data Protection Directive of 1995 and is intended to strengthen individuals' fundamental rights in the digital age. Concerning the purpose of the GDPR, and to reflect on its *democratic accountability* dimension, I ask whether accountability mechanisms accord with substantive standards and democratically legitimized preferences. It is possible to identify two distinct aims of the GDPR. On the one hand, when tabling the proposal for the GDPR in 2012, the Commission stated that an important aim of the regulation is to support its digital single market (e.g. Recitals 2, 7 and 13, GDPR). It argued that "building trust in the online environment is key to economic development" (European Commission, 2012). In this regard, the GDPR was a continuation of the necessity to create consumers' trust in the digital economy for facilitating the development of the EU digital single market. Transforming the 1995 Directive into a regulation was meant to facilitate business development by clarifying rules for companies and public bodies in the digital single market, thereby addressing also the problem of fragmentation along national borders and unnecessary administrative burdens. The creation of a single market, and the role of the GDPR in facilitating this project, can be regarded as one of the values underlying the European integration project because market integration (Article 3(3) TEU) has always been meant to promote European integration thereby serving such goals as ultimately the prevention of war.

On the other hand, the GDPR is set to be designed to strengthen individuals' fundamental rights in the digital age, in accord with core European democratic values. Building upon the 1995 data protection directive, data protection has also become part of the EU's Charter of Fundamental Rights (CFR) proclaimed in 2000, which is legally binding since the entry into force of the Lisbon Treaty in 2009 and laid out a strong Treaty basis for a firm EU data protection regulation in form of the GDPR. In accordance with its Recital 1, the GDPR stresses that the protection of natural persons in relation to the processing of personal data is a fundamental right. This is laid out, inter alia, in Article 8(1) of the Charter of Fundamental Rights of the European Union and Article 16(1) of the Treaty on the Functioning of the European Union (TFEU), which provide that everyone has the right to the protection of personal data concerning them. One example of its application is the so called "right to be forgotten". Article 17 of the GDPR enshrines that citizens have the right to request the erasure or removal of inaccurate data. In this regard, the GDPR accords with one of the EU's core substantive standards as they are codified in the CFR.

## Constitutional Accountability of the GDPR

Regarding the constitutional accountability dimension, I first ask if the accountability mechanisms of the GDPR allow for monitoring and curtailing the abuse of powers and privilege. In this regard, it is noteworthy that the GDPR relies less on private authority, as it was practised in many areas concerning digital market integration (Farrand, 2023), exercised for example through business trust marks. Already the 1995 Data Protection Directive required member states to set up independent national Data Protection Authorities (DPAs), which are also responsible for monitoring the application and enforcing the GDPR (Article 51 – 59). DPAs are given complete independence in performing their tasks and exercising their powers (Article 52(1) GDPR). This applies also to the members of the supervisory authority, whom the text, among other things, designs to be appointed in a transparent procedure by democratically legitimate bodies (such as national governments or parliaments, Article 53 GDPR). They may be dismissed only in cases of serious misconduct, which is meant to also secure personal independence of the supervisory authority. These national data protection authorities cooperate in the European Data Protection Board (EDPB, Article 60 – 76 GDPR). The EDPB is a body of the EU with own legal personality. Other EU institutions, notably the Commission, have the right to participate in the activities and meetings of the Board, but without voting rights. In this regard, the GDPR sets up a member state level structure with cooperative elements on the supranational level to ensure coherent application.

Constitutional accountability requires further assessing if public institutions possess credible enforcement mechanisms. Enforcement of the GDPR lies with the national DPAs (Article 57 GDPR). They possess relatively comprehensive investigative and corrective powers (Article 58 GDPR). Among other things, the national DPAs can carry out investigations, order the rectification or erasure of personal data, and impose administrative fines. For severe violations, the fine framework can be up to 20 million euros, or up to 4 percent of total company global turnover, whichever is higher. Also, the catalogue of less severe violations provides for fines of up to EUR 10 million, or 2 percent of global turnover (Article 83(4-6) GDPR). While these amounts appear to be fairly comprehensive, enforcement lies with the national DPAs, which have relatively high discretion in this area and whose enforcement practices differ substantially (Sivan-Sevilla, 2022).

The enforcement structure of the GDPR relies exclusively on member state authorities. Supranational bodies, like for example the Commission, have no formal role in the enforcement of the GDPR. Also, the newly created EDPB has a "mere" coordinative role and issues for example guidelines and recommendations (Article 68-76, GDPR). This exclusive reliance on member state authority continues the country of origin principle which was often used in the context of legislation of the EU single market. The argument is that when companies are supervised by the authorities in the country in which they are established and member states mutually recognise the authority of each other's DPAs, legal fragmentation can be prevented and companies and citizens can benefit from a frictionless internal market.

However, the used country of origin principle creates supervisory bottlenecks. It is the Irish data protection authority that is in charge of supervising most Big Tech companies because Facebook, Twitter, Google, and Apple have their headquarters in Ireland. This is problematic because, according to a report published by the Irish Council for Civil Liberties (Ryan and Toner, 2021), an NGO, the Irish authority, given a lack of resources or of political will, has proven unable to police Big Tech firms' use of personal data and is accused of paralysing the regulation. In this regard, the GDPR shows weaknesses in the constitutional dimension of platform accountability.

## Platform Accountability in the Digital Services Act

### Strengthening Democratic Accountability through the DSA

The DSA is one of two regulations of the EU's Digital Services Package that has been drafted by the Commission in late 2020 and adopted in first reading by the EP in July 2022. According to the Commission, the DSA is intended to rebalance "the responsibilities of users, platforms, and public authorities [...] according to European values, placing citizens at the centre" (European

Commission, 2022). According to the Act, this is important because the "digital transformation [...] resulted in new risks and challenges for individual recipients [...], companies and society as a whole" (Recital 1, DSA). Its key aims are better protection of consumers and their fundamental rights online, more powerful transparency and an accountability framework for online platforms, as well as fostering innovation, growth, and competitiveness within the single market (European Commission, 2022). In this regard, the aims set out in the DSA do reflect core EU democratic standards. It reflects the EU's dual commitment to the establishment of a competitive internal market (Article 3(3) TEU) and the aim of protecting core democratic values such as the freedom of expression online (Article 11, CFR).

To achieve these goals, the DSA maintains more market-liberal principles like the avoidance of general monitoring of users' online activities on platforms (Article 8, DSA) and a reliance on platform self-regulation in the form of codes of conduct and standards of best practice (Farrand, 2023). On the other hand, new measures are designed to counter illegal goods, services, or content online, by establishing a mechanism for users to flag such content. Furthermore, a ban on certain types of targeted advertisements on online platforms was adopted, specifically when they target children (Article 28 DSA) or when they use special categories of personal data (Article 26(3) DSA), like racial or ethnic origin, religion or sexual orientation, in accordance with the GDPR (Article 9(1)).

## Strengthening Constitutional Accountability through the DSA

It is important to note that some aspects of the DSA continue the previously used reliance on "regulated self-regulation" (Farrand, 2023) and stakeholder involvement, rather than in form of external public supervision. One example is the fight against illegal content. Inter alia, platforms will keep using so called trusted flaggers, which are entities with certain expertise in a field that are aptly targeting illegal content. Platforms are obliged to ensure that notices submitted by trusted flaggers are given priority and are processed and decided upon without undue delay (Article 22 DSA). Another example would be risk assessments and the mitigation of risks. In these areas, very large online platforms are expected to assess systemic risks stemming from the design or functioning of their services (Article 34 DSA) and they are obliged to put in place mitigation measures, tailored to the specific systemic risks identified (Article 35 DSA).

It is important to note, too, that these measures are designed, implemented and reported by the platforms themselves. However, the DSA also strengthens public accountability by implementing tougher oversight through a new system of Digital Services Coordinators. These are institutions at member state level and tasked with applying and enforcing the DSA (Article 49 DSA). Here the DSA departs from the principle of regulated self-regulation, and Digital Services Coordinators receive powers in that they award the status of trusted flaggers. Furthermore, self-regulation regarding systemic risks of very large online platforms requires external audits (Article 37 DSA), conducted by independent audit organisations without connections to the platform company. Platforms choose and pay audit organizations which may lead to audit capture (Laux, Wachter and Mittelstadt, 2021), that is unjustified positive reviews based on close business relationships and not on objective evaluations. However, Digital Services Coordinators receive the authority to require auditors to provide information that relates to suspected infringement of the DSA. In this regard, the DSA improves public monitoring and the curtailing of the abuse of powers and privilege.

The design of the supervisory structure of the DSA shows that the regulation learns lessons from the weaknesses of the GDPR and improves both aspects of constitutional accountability, monitoring and enforcement. In particular, the DSA improves the enforcement problem known from the GDPR (see above) by distinguishing between different categories of online intermediaries regarding their role, size and impact. The more important and bigger the platforms are, the more obligations apply. The most obligations apply to very large online platforms, defined as those with 45 million or more monthly users (Article 33 DSA). These very large online platforms are supervised directly by the Commission (Article 65 DSA), thereby preventing member state level enforcement bottlenecks known from the GDPR and ensuring the level playing field of the EU digital single market. Both, Digital Services Coordinators and the Commission can impose fines for misconduct, which can amount to up to six percent of the company's total worldwide annual turnover, which is two percent higher than in severe cases defined in the GDPR. However, as regards implementation and enforcement, the Commission has to prove that it can make difficult choices that require balancing competing goods such as the freedom of expression and the prevention of online harms.

## Platform Accountability in the Digital Markets Act

### Democratic Accountability through the DMA

The DMA is the second regulation of the EU's Digital Services Package. In the context of the DMA, the first evaluative question is mostly answered in terms of ensuring fair competition and a level playing field for platforms, businesses and users. According to Recital 2, adjusted competition rules are necessary because platform companies' characteristics can be exploited, for example because of their extreme scale economies that lead to nearly zero marginal costs to add business users or end users. This can create lock-in effects and lead to a concentration of power.

The DMA particularly addresses so called gatekeepers (see below). For these gatekeepers, the DMA defines a full set of do's and don'ts. For example, gatekeepers must allow third parties to interoperate with the gatekeeper's own services (Article 7, DMA) and are prohibited from treating services and products offered by the gatekeeper itself more favourably in ranking than similar services or products offered by third parties (Article 6(5), DMA). The approach of defining a clear set of do's and don'ts changes the EU's traditional competition policy approach. Previously, competition policy worked ex-post, meaning that intervention occurred after an infringement happened, which was typically proven by higher prices at the expense of users. The DMA's ex-ante approach promotes democratic accountability predominantly in its economic dimension. It addresses the market structure of the digital economy, and allows regulators to act before one of the EU's goals, namely the establishment of a competitive internal market (Article 3(3) TEU), is levered.

### Strengthened Constitutional Accountability through the DMA

From a perspective of constitutional accountability, in line with my first evaluative question, it is important to note that the DMA deals with so-called gatekeepers only. They are defined according to a number of qualitative and quantitative criteria and their designation allows further discretion of the Commission. In particular, gatekeepers are defined as those companies with a significant impact on the internal market, those providers of important gateway for business users to reach end users, and those that hold this with an entrenched and durable position (Article 3(1) DMA). More quantitative criteria include inter alia whether companies achieve an annual turnover of more than EUR 7,5 billion and if the platform has 45 million or more monthly users (Article 3 (2), DMA). Moreover, the Commission has the power to designate gatekeepers on the basis of a comprehensive catalogue of additional criteria (Article 3(8), DMA). These criteria promote constitutional platform accountability by allowing for focussed monitoring, and shall prevent overburdening supervisors.

To ensure effective enforcement, the DMA delegates this mandate to the Commission, which corresponds with its tasks in competition policy more generally. The Commission receives relatively far-reaching powers, like requesting all necessary information (Article 21, DMA), conducting inspections (Article 23, DMA) and issuing fines (Article 30, DMA). To ensure credibility, the DMA foresees sanctions of up to 10 percent of the company's total worldwide annual turnover (Article 30(1), DMA) or even 20 percent in the event of repeated infringements (Article 30(2), DMA). For enforcing the DMA, the responsible Directorate General (DG) will not be the experienced DG Competition, dealing with the enforcement of competition policy, but the DG for Communications Networks, Content and Technology (Martins and Carugati, 2022). The Commission aims to establish structures for effective cooperation between the two DGs. By the time of writing in the summer of 2023, the effective implementation of the DMA remains to be seen. The regulation intends strengthening platform accountability, in particular in its constitutional dimension.

## Conclusion

Given their ever-increasing powers, Big Tech companies have developed functions and acquired competences that were previously the prerogative of the nation state. In response, the European Commission issued a number of legislative proposals to curtail their powers and to align the responsibilities of platforms with European values, to strengthen the competences of public authorities, thereby increase the framework through which the public holds large platform companies accountable. By analysing three far-reaching regulations, the GDPR, the DSA and the DMA, this article has shown what the Commission's attempts mean in actual regulation. For this purpose, I distinguished between a more democratic and more constitutional dimension of what I call a public form of platform accountability. This allowed me to show through which instruments the Commission aims to hold platform companies to account, and to assess different qualities of the platform accountability framework that is emerging in the EU.

My analyses contributed to the literature on platform accountability by revealing strengths and weaknesses in the more democratic and the more constitutional dimensions of accountability of the three regulations. In particular, the GDPR has improved the dimension of democratic accountability by aligning the practices of digital service providers with EU fundamental rights. However, my analysis has also revealed weaknesses in the more constitutional dimensions of accountability, particularly concerning the enforcement of the GDPR. By setting rules for content moderation in line with European values, the DSA enhances the democratic accountability of platform companies in the EU. It further reflects lessons from the weaknesses of the GDPR and improves the dimension of constitutional accountability by aiming for stronger enforcement through the delegation of supervisory powers over very large online platforms to the Commission. Similar to classical competition law, the DMA enhances mostly the prevention of abuse of power and privilege by very large online platforms, and it establishes new ex-ante instruments to ensure competitive markets.

It is important to highlight that by controlling the digital sphere, a few Big Tech companies have acquired substantial socio-political power over ever-increasing parts of life in the 21st century. To exercise what many in the EU now call digital sovereignty (Falkner et al., 2022), it is crucial to establish platform public accountability, which shall ultimately allow cooperation for tackling pressing global problems like making societies fairer and developing more sustainably (see also Floridi, 2020, p. 374). Analysing the three influential regulations separately has enabled me to identify strengths and weaknesses in their accountability dimensions. Together, they can provide a robust and synergistic framework for platform accountability in the EU. In particular, mostly the GDPR and DSA reinforce the democratic dimension of platform accountability in the EU, while the DMA targets the more constitutional dimension by addressing concentration and abuse of power. On the constitutional dimension, I have also found that the DSA has been improved on the basis of the GDPR experience by delegating platform oversight to the EU Commission to ensure effective and coherent enforcement of the regulation, similar to the powers it has in competition policy and the DMA.

By demonstrating that the EU aims at strengthening mechanisms of platform accountability, I have also identified weak spots. While the approach to empower the Commission is in principle sensible, given its supranational position that can ensure coherent application across the single market, the approach brings potential challenges. For example, the DSA is legislation dealing with politically sensitive issues in terms of content moderation and might have to balance partly contradictory goals like ensuring the freedom of expression and preventing online harm through the spread of disinformation and hate speech. The Commission is, however, not an apolitical body, so its decisions could be regarded as biased. The Commission will have to prove its political sensitivity when enforcing the DSA to remain a credible platform supervisor. Furthermore, the new powers will require sufficient resources, which are currently mainly available to national authorities. Thus, the Commission will also have to ensure cooperation without loss of efficiency.

## Bibliography

**Abrams, Abigail** (2019), "Here's What We Know So Far About Russia's 2016 Meddling", *Time*, 18 April, <https://time.com/5565991/russia-influence-2016-election/>.

**Bartoš, Ivan** (2022), "DMA: Council gives final approval to new rules for fair competition online" [Press release], 18 July, <https://www.consilium.europa.eu/en/press/press-releases/2022/07/18/dma-council-gives-final-approval-to-new-rules-for-fair-competition-online/>.

**Bertuzzi, Luca** (2021), "Leading MEP's bid to 'take back control of Big Tech'", *Euractiv*, 21 Juni, <https://www.euractiv.com/section/digital/news/leading-meps-bid-to-take-back-control-of-big-tech/>.

**Bovens, Mark** (2005), "Public Accountability", *in* Ewan Ferlie, Laurence E. Lynn Jr. and Christopher Pollitt (eds), *The Oxford Handbook of Public Management*, Oxford, Oxford University Press, p. 182-208.

**Bovens, Mark** (2010), "Two Concepts of Accountability: Accountability as a Virtue and as a Mechanism", *West European Politics*, vol. 33 (5), p. 946-967, <doi:10.1080/01402382.2010.486119>.

**Bovens, Mark, Schillemans, Thomas and Goodin, Robert** (2014), "Public Accountability", *in* Mark Bovens, Thomas Schillemans and Robert Goodin (eds), *The Oxford Handbook of Public Accountability*, Oxford, Oxford University Press, p. 0, <doi:10.1093/oxfordhb/9780199641253.013.0012>.

**Bovens, Mark, Schillemans, Thomas and 'T Hart, Paul** (2008), "Does Public Accountability Work? An Assessment Tool", *Public Administration*, vol. 86 (1), p. 225-242, <https://doi.org/10.1111/j.1467-9299.2008.00716.x>.

**Bradford, Anu** (2020), *The Brussels Effect: How the European Union Rules the World*, New York, Oxford University Press, <doi:10.1093/oso/9780190088583.001.0001>.

**Chen, Adrian** (2018), "Cambridge Analytica and Our Lives Inside the Surveillance Machine", *The New Yorker*, 21 March, <https://www.newyorker.com/tech/annals-of-technology/cambridge-analytica-and-our-lives-inside-the-surveillance-machine>.

**Cini, Michelle and Czulno, Patryk** (2022), "Digital Single Market and the EU Competition Regime: An Explanation of Policy Change", *Journal of European Integration*, vol. 44 (1), p. 41-57, <doi:10.1080/07036337.2021.2011260>.

**Cioffi, John W., Kenney, Martin F. and Zysman, John** (2022), "Platform power and regulatory politics: Polanyi for the twenty-first century", *New Political Economy*, vol. 27 (5), p. 820-836, <doi:10.1080/13563467.2022.2027355>.

**Damro, Chad** (2012), "Market power Europe", *Journal of European Public Policy*, vol. 19 (5), p. 682-699, <doi:10.1080/13501763.2011.646779>.

**European Commission** (2012), *Proposal for a regulation on the protection of individuals with regard to the processing of personal data and on the free movement of such data (General Data Protection Regulation)*, COM(2012) 11 final, Brussels, 25 January.

**European Commission** (2015), "A Digital Single Market Strategy for Europe", COM(2015) 192 final, Brussels, 6 May.

**European Commission** (2016), "Communication on Online Platforms and the Digital Single Market Opportunities and Challenges for Europe", COM(2016) 288 final, Brussels, 25 May.

**European Commission** (2022), "The Digital Services Act: ensuring a safe and accountable online environment", *European Commission*, <https://ec.europa.eu/info/strategy/priorities-2019-2024/europe-fit-digital-age/digital-services-act-ensuring-safe-and-accountable-online-environment_en>.

**European Council** (1994), "Europe and the global information society. Recommendations of the high-level group on the information society to the Corfu European Council (Bangemann group)", *Bulletin of the European Union*, Supplement 2/94, Luxembourg, Office for the Official Publications of the European Communities, p. 5-41.

**Falkner, Gerda, Heidebrecht, Sebastian, Obendiek, Anke and Seidl, Timo** (2022), "Digital Sovereignty - Rhetoric and Reality", Framework paper for the Online Conference, Centre for European Integration Research, University of Vienna, 28-29 April.

**Farrand, Benjamin** (2023), "The Ordoliberal Internet? Continuity and Change in the EU's approach to the Governance of Cyberspace", *European Law Open*, vol. 2 (1), p. 106-127.

**Floridi, Luciano** (2020), "The Fight for Digital Sovereignty: What It Is, and Why It Matters, Especially for the EU", *Philosophy & Technology*, vol. 33 (3), p. 369-378, <doi:10.1007/s13347-020-00423-6>.

**Gillespie, Tarleton** (2018), "Regulation of and by Platforms", *in* Jean Burgess, Alice Marwick and Thomas Poell (eds), *The SAGE Handbook of Social Media*, London, Sage Publications, p. 254-278, <doi:10.4135/9781473984066>.

**Gorwa, Robert** (2019), "What is platform governance?", *Information, Communication & Society*, vol. 22 (6), p. 854-871, <doi:10.1080/1369118X.2019.1573914>.

**Haggart, Blayne and Keller, Clara Iglesias** (2021), "Democratic legitimacy in global platform governance", *Telecommunications Policy*, vol. 45 (6), p. 102-152, <doi:https://doi.org/10.1016/j.telpol.2021.102152>.

**Heidebrecht, Sebastian** (2022), "Mapping continuity and change of goals, protagonists and instruments of EU digital single market governance over three decades", *EIF Working Paper*, vol. 1, <https://eif.univie.ac.at/downloads/workingpapers/wp2022-01.pdf>.

**Heidebrecht, Sebastian** (2023), "From Market Liberalism to Public Intervention: Digital Sovereignty and Changing European Union Digital Single Market Governance", *Journal of Common Market Studies*, <doi:https://doi.org/10.1111/jcms.13488>.

**Helberger, Natali** (2020), "The Political Power of Platforms: How Current Attempts to Regulate Misinformation Amplify Opinion Power", *Digital Journalism*, vol. 8 (6), p. 842-854, <doi:10.1080/21670811.2020.1773888>.

**Helberger, Natali, Pierson, Jo and Poell, Thomas** (2018), "Governing online platforms: From contested to cooperative responsibility", *The Information Society*, vol. 34 (1), p. 1-14, <doi:10.1080/01972243.2017.1391913>.

**Khan, Lina M.** (2017), "Amazon's antitrust paradox", *Yale Law Journal*, vol. 126 (3), p. 712–805.

**Klonick, Kate** (2018), "The New Governors: The People, Rules, and Processes Governing Online Speech", *Harvard Law Review*, vol. 131 (6), p. 1598-1670.

**Klonick, Kate** (2019), "Does Facebook's Oversight Board Finally Solve the Problem of Online Speech?", *Centre for International Governance Innovation*, 28 October, p. 51-53.

**Laurer, Moritz and Seidl, Timo** (2021), "Regulating the European Data-Driven Economy: A Case Study on the General Data Protection Regulation", *Policy & Internet*, vol. 13 (2), p. 257-277, <doi:10.1002/poi3.246>.

**Laux, Johann, Wachter, Sandra and Mittelstadt, Brent** (2021), "Taming the few: Platform regulation, independent audits, and the risks of capture created by the DMA and DSA", *Computer Law & Security Review*, vol. 43, p. 105-613, <doi:https://doi.org/10.1016/j.clsr.2021.105613>.

**Lehdonvirta, Vili** (2022), *Cloud Empires: How Digital Platforms Are Overtaking the State and How We Can Regain Control*, Cambridge, Mass., MIT Press.

**Manners, Ian** (2002), "Normative Power Europe: A Contradiction in Terms?", *Journal of Common Market Studies*, vol. 40 (2), p. 235-258, <doi:https://doi.org/10.1111/1468-5965.00353>.

**Martins, Catarina and Carugati, Christophe** (2022), "Insights for successful enforcement of Europe's Digital Markets Act", *Bruegel Blog*, 11 May, <https://www.bruegel.org/blog-post/insights-successful-enforcement-europes-digital-markets-act>.

**Mazzetti, Mark and Schmidt, Michael S.** (2013), "Ex-Worker at C.I.A. Says He Leaked Data on Surveillance", *The New York Times*, 9 June, <https://www.nytimes.com/2013/06/10/us/former-cia-worker-says-he-leaked-surveillance-data.html>.

**Meunier, Sophie and Mickus, Justinas** (2020), "Sizing up the competition: explaining reform of European Union competition policy in the Covid-19 era", *Journal of European Integration*, vol. 42 (8), p. 1077-1094, <doi:10.1080/07036337.2020.1852232>.

**Newman, Abraham L.** (2008), "Building Transnational Civil Liberties: Transgovernmental Entrepreneurs and the European Data Privacy Directive", *International Organization*, vol. 62 (1), p. 103-130, <doi:10.1017/S0020818308080041>.

**Newman, Abraham L.** (2020), "Digital Policy-Making in the European Union. Building the New Economy of an Information Society", *in* Helen Wallace, Mark Pollack, Christilla Roederer-Rynning and Alasdair Young (eds), *Policy-Making in the European Union*, 8th edition, Oxford, Oxford University Press, p. 276-296.

**Nieborg, David B. and Poell, Thomas** (2018), "The platformization of cultural production: Theorizing the contingent cultural commodity", *New Media & Society*, vol. 20 (11), p. 4275-4292, <doi:10.1177/1461444818769694>.

**Obendiek, Anke** (2021), "Take back control? Digitale Souveränität und Europas digitale Zukunft", *Hertie School Jacques Delors Centre Policy Paper*, 12 May.

**O'Donnell, Guillermo A.** (1998), "Horizontal Accountability in New Democracies", *Journal of Democracy*, vol. 9 (3), p. 112-126.

**Pasquale, Frank A.** (2016), "Two narratives of platform capitalism", *Yale Law & Policy Review*, vol. 35 (1), p. 309–319.

**Pasquale, Frank A.** (2018), "Tech platforms and the knowledge problem", *American Affairs*, vol. 2 (2), p. 3-16.

**Pistor, Katharina** (2020), "Rule by Data: The End of Markets?", *Law and Contemporary Problems*, vol. 83, p. 101-124.

**Plucinska, Joanna** (2018), "EU tries to shield 2019 election from fake news", *Politico*, 25 April, <https://www.politico.eu/article/fake-news-eu-parliament-election-commission-gathers-national-election-officials-as-ballot-fears-rise/>.

**Rahman, K. Sabeel** (2017), "From Civic Tech to Civic Capacity: The Case of Citizen Audits", *Political Science & Politics*, vol. 50 (3), p. 751-757, <doi:10.1017/S1049096517000543>.

**Ryan, Johnny and Toner, Alan** (2021), *Europe's enforcement paralysis. ICCL's 2021 report on the enforcement capacity of data protection authorities*, Irish Council for Civic Liberties, <https://www.iccl.ie/digital-data/2021-gdpr-report/>.

**Scharpf, Fritz W.** (1999), *Governing in Europe: Effective and democratic?*, Oxford, Oxford University Press.

**Schedler, Andreas** (1999), "Conceptualizing Accountability", *in* Andreas Schedler, Larry J. Diamond and Marc F. Plattner (eds), *The Self-restraining State: Power and Accountability in New Democracies*, Boulder, Lynne Rienner, p. 13-28.

**Schmidt, Vivien A.** (2013), "Democracy and Legitimacy in the European Union Revisited: Input, Output and 'Throughput'", *Political Studies*, vol. 61 (1), p. 2-22, <doi:https://doi.org/10.1111/j.1467-9248.2012.00962.x>.

**Schmitter, Philippe C.** (2004), "The Quality of Democracy: The Ambiguous Virtues of Accountability", *Journal of Democracy*, vol. 15 (4), p. 47-60

**Schmitz, Luuk and Seidl, Timo** (2022), "As Open as Possible, as Autonomous as Necessary: Understanding the Rise of Open Strategic Autonomy in EU Trade Policy", *Journal of Common Market Studies*, vol. 61 (3), p. 834-852, <doi:https://doi.org/10.1111/jcms.13428>.

**Schünemann, Wolf Jürgen and Windwehr, Jana** (2021), "Towards a 'gold standard for the world'? The European General Data Protection Regulation between supranational and national norm entrepreneurship", *Journal of European Integration*, vol. 43 (7), p. 859-874, <doi:10.1080/07036337.2020.1846032>.

**Seawright, Jason and Gerring, John** (2008), "Case Selection Techniques in Case Study Research: A Menu of Qualitative and Quantitative Options", *Political Research Quarterly*, vol. 61 (2), p. 294-308, <doi:10.1177/1065912907313077>.

**Sivan-Sevilla, Ido** (2022), "Varieties of enforcement strategies post-GDPR: a fuzzy-set qualitative comparative analysis (fsQCA) across data protection authorities", *Journal of European Public Policy*, p. 1-34, <doi:10.1080/13501763.2022.2147578>.

**Suzor, Nicolas P.** (2019), *Lawless. The Secret Rules That Govern our Digital Lives*, Cambridge, Cambridge University Press.

**Tucker, Joshua A., Theocharis, Yannis, Roberts, Margaret E. and Barberá, Pablo** (2017), "From Liberation to Turmoil: Social Media And Democracy", *Journal of Democracy*, vol. 28 (4), p. 46-59.

**Vaidhyanathan, Siva** (2018), *Antisocial media: How Facebook disconnects us and undermines democracy*, Oxford, Oxford University Press.

**Valdez, Jimena** (2023), "The politics of Uber: Infrastructural power in the United States and Europe", *Regulation & Governance*, vol. 17 (1), p. 177-194, <doi:https://doi.org/10.1111/rego.12456>.

**Van Dijck, José, Nieborg, David B. and Poell, Thomas** (2019), "Reframing platform power", *Internet Policy Review*, vol. 8 (2), <doi:10.14763/2019.2.1414>.

**Van Doorn, Niels** (2017), "Platform labor: on the gendered and racialized exploitation of low-income service work in the 'on-demand' economy", *Information, Communication & Society*, vol. 20 (6), p. 898-914, <doi:10.1080/1369118X.2017.1294194>.

**Vestager, Margrethe** (2022), "Digital Services Act: Commission welcomes political agreement on rules ensuring a safe and accountable online environment" [Press release], Brussels, 23 April, <https://ec.europa.eu/commission/presscorner/detail/en/IP_22_2545>.

**Wong, David and Floridi, Luciano** (2022), "Meta's Oversight Board: A Review and Critical Assessment", *Minds and Machines*, vol. 33, p. 261-284, <doi:10.1007/s11023-022-09613-x>.

**Wu, Tim** (2018), *The Curse of Bigness: Antitrust in the New Gilded Age*, New York, Columbia Global Reports.

**Zuboff, Shoshana** (2019), *The age of surveillance capitalism. The fight for a human future at the new frontier of power*, New York, Hachette Book Group.

**Dr. Sebastian Heidebrecht** is a postdoctoral researcher at the Centre for European Integration Research (EIF); Institute of Political Science, University of Vienna. He is co-editor of a special issue on digital sovereignty in the Journal of European Public Policy (JEPP) and recently published on the Digital Single Market in the Journal of Common Market Studies (JCMS).

sebastian.heidebrecht@univie.ac.at

# Lectures critiques

POLITIQUE EUROPÉENNE
N° 81 | 2023
Julien Louis
p. 170-175

Aurélie Dianara Andry, *Social Europe, the Road not Taken. The Left and European Integration in the Long 1970s*, Oxford, Oxford University Press, 2022, 333 pages.

**Julien Louis**
Université Lyon 2

Tiré d'une thèse d'histoire soutenue à l'Institut universitaire européen en 2017, l'ouvrage d'Aurélie Dianara Andry prend pour objet les projets d'Europe sociale portés par la gauche politique et syndicale au cours de la « longue » décennie 1970. En partant de la fin des années 1960 pour s'arrêter au début des années 1980, la périodisation retenue est en soi une première contribution originale de cette recherche, car elle permet un regard neuf et rafraichissant sur cet objet – pourtant saturé politiquement et scientifiquement – qu'est l'Europe sociale. L'autrice prend en effet le contrepied de nombreux auteurs qui voient dans la Présidence de la Commission européenne de Jacques Delors (1985-1995) un « âge d'or » de l'Europe sociale.

Ce livre montre qu'il n'en est rien : l'acmé des projets d'Europe sociale se situe plutôt dans les années 1970, car c'est alors que la gauche politique et syndicale défendit les propositions sociales les plus audacieuses de l'histoire de l'intégration européenne : redistribution des richesses, régulation du marché, planification économique, contrôle public des investissements et des entreprises (en particulier des multinationales), démocratisation de l'économie, harmonisation par le haut des régimes sociaux et fiscaux en Europe, et même protection de l'environnement ainsi que rééquilibrage des relations internationales en faveur des pays du Sud. Une telle « route » ne fut toutefois pas empruntée, ainsi que l'annonce d'emblée le titre de l'ouvrage. Mais plus encore qu'un possible non advenu, il s'agit d'un possible oublié, dont l'échec fut masqué par le néolibéralisme triomphant des années 1980 et les quelques mesures sociales mises en place après Maastricht. L'autrice montre ainsi que ces dernières constituent en réalité un renoncement majeur par rapport aux ambitions réformatrices des années 1970.

Pour exhumer ces « programmes forts » oubliés de l'Europe sociale, Aurélie Dianara Andry propose une analyse historique qui prend pour objet à la fois le contenu de ces programmes (contradictoires, flous, évolutifs) et les acteurs (tout aussi divers et même divisés) qui les portent. Concrètement, l'analyse historique fait tenir ensemble les projets d'Europe sociale défendus par les socialistes, les communistes, et les syndicats, ceci sans jamais donner une image homogène de la gauche européenne, qu'il faudrait en toute rigueur ne nommer qu'au pluriel. Sans qu'elle s'en revendique explicitement, l'autrice propose ainsi une histoire sociale rigoureuse des idées politiques, attentive aux contextes d'énonciation des projets d'Europe sociale, aux acteurs qui les formulent, à leurs modes d'organisation collective, à leurs mobilisations pour les faire advenir, et enfin aux raisons de leurs échecs.

La perspective adoptée est résolument celle d'une histoire transnationale de l'Europe sociale, puisque la focale est mise sur les débats qui animent les espaces transnationaux construits ou investis par la gauche : l'Internationale socialiste, le Parlement européen et ses groupes, la Confédération des Partis socialistes de la CE (CPSCE), la Confédération européenne des syndicats (CES), les sommets et les congrès européens. Cette histoire transnationale est en outre solidement adossée à une histoire comparée des gauches européennes et de leurs rapports à l'Europe, en particulier françaises, allemandes, anglaises et italiennes, et dans une moindre mesure, néerlandaises, belges et scandinaves. On regrettera à ce propos qu'aucune réflexion ne soit menée sur la constitution du corpus d'archives et des sources de seconde main nécessaires pour mener à bien une telle analyse transnationale et comparée de l'Europe sociale. Les fonds mobilisés n'apparaissent qu'en notes de bas page, sans même une annexe dédiée.

Au total, l'ouvrage s'organise en six chapitres rédigés dans une écriture dont la mise en page (quelques rares titres dans les chapitres, pas d'espaces entre les paragraphes...) ainsi que la densité requièrent une lecture exigeante. Cette densité constitue toutefois la contrepartie d'un souci d'exhaustivité dans la restitution des éléments empiriques nécessaires à la compréhension fine des configurations transnationales dans lesquelles la gauche politique et syndicale énonce et défend ses revendications successives en faveur de l'Europe sociale.

Le premier chapitre (« The 'Social Dimension' of Postwar Europe ») s'attache à contextualiser les enjeux de l'Europe sociale en revenant sur le déficit social initial de l'intégration européenne. Si les socialistes et les syndicalistes sont impliqués dans la construction européenne dès ses origines, et si le sens de celle-ci n'a jamais été ni univoque ni prédéterminé, il apparaît assez clairement que les conservateurs, les démocrates-chrétiens et les néo-libéraux constituent les forces politiques dominantes lors des traités de Paris et de Rome. Qui plus est, la fondation des Communautés est en partie motivée par des dynamiques atlantiques dans le contexte de la guerre froide, ce que les syndicats et partis communistes dénonçaient alors vivement. Faibles, les compétences sociales déléguées aux autorités communautaires sont également oblitérées par la croyance que l'essor économique permis par le marché commun entraînera mécaniquement un progrès social. En dépit de la faiblesse originelle de l'Europe sociale, l'autrice souligne que tout n'était alors pas joué : la route pouvait encore être ouverte.

Le deuxième chapitre (« 1968, 1969 : Social Protest, European 'Revival' ») traite de la puissante vague de conflictualité sociale qui parcourt l'Europe à partir de la fin des années 1960 et les effets que ce cycle de mobilisations produit sur le traitement de la question sociale par les Communautés. Largement négligée jusqu'alors, l'Europe sociale est mise au plus haut niveau de l'agenda communautaire en 1969, lors du sommet de La Haye qui se tient les 1$^{er}$ et 2 décembre. Outre la multiplication des mouvements sociaux, cette mise à l'agenda est rendue possible par l'arrivée au pouvoir de Willy Brandt en Allemagne de l'Ouest, concomitante de la démission de Charles de Gaulle, mais aussi par l'intensification de la coopération entre les gauches européennes, au Parlement européen notamment, qui se traduit par la création de structures partisanes et syndicales européennes, aux ressources et au pouvoir encore modestes toutefois. Ces forces de gauche se mobilisent pour peser sur le sommet de La Haye : si les résultats ne sont pas à la hauteur de ceux qui étaient espérés, les chefs d'État et de gouvernement s'accordent sur la nécessité de renforcer la politique sociale européenne.

Le troisième chapitre (« A New Social Wind ») analyse les conséquences du sommet de La Haye pour l'Europe sociale. Les engagements qui sont pris alors demeurent vagues, leurs effets sont limités (création du Comité permanent pour l'emploi, réforme du Fonds social européen) et tout l'enjeu consiste à les concrétiser. La décennie 1970 semble justement s'ouvrir sous des auspices favorables pour la gauche européenne : celle-ci accède au pouvoir dans plusieurs États membres ou en voie d'adhésion à la CE, le socialiste belge Raymond Rifflet est nommé à la tête de la DG Affaires sociales de la Commission européenne, tandis que le Néerlandais Sicco Mansholt, également social-démocrate, en devient le président. Cette conjoncture propice à l'Europe sociale permet des avancées lors du sommet de Paris, qui se tient les 19-21 octobre 1972. Sous l'impulsion de Willy Brandt, qui prône une « Union sociale » comme le nécessaire pendant de l'Union économique et monétaire, les chefs d'État et de gouvernement chargent ainsi la Commission d'élaborer un programme d'action sociale pour janvier 1974.

Le quatrième chapitre (« 'For a Social Europe' ») se concentre sur l'élaboration de ce programme d'action sociale et sur les tentatives de la gauche pour en influencer le contenu. Cette dernière doit donner un contenu précis à l'idée d'Europe sociale, ce qui l'oblige à amorcer un travail programmatique. Ce travail est d'abord fait par les syndicats, dont la structuration européenne prend une nouvelle ampleur avec la création de la CES en février 1973. Les 26 et 27 avril de cette même année se tient également le 9e Congrès des partis socialistes européens à Bonn, qui voit l'adoption des « thèses pour une Europe sociale », articulées en trois grands objectifs : la planification de l'économie pour réaliser le plein emploi, l'amélioration des conditions de vie et de travail, et la convergence vers le haut des systèmes de protection sociale. Malgré ces revendications communes, le congrès met au jour les divisions qui traversent la social-démocratie européenne, par exemple sur le degré de participation des travailleurs à la vie de l'entreprise. Finalement, le résultat n'est pas à la hauteur des efforts fournis, puisque les propositions de la Commission en octobre 1973, et encore plus celles actées par le Conseil en janvier 1974, apparaissent très éloignées des revendications élaborées par la gauche.

Le cinquième chapitre (« There Were Alternatives ») étudie les tentatives de la gauche européenne pour formuler une réponse économique coordonnée à la crise après le choc pétrolier de 1973. C'est à cette période que l'influence des mouvements sociaux se fait le plus sentir, ceux-ci tirant vers la gauche une partie des sociaux-démocrates européens, comme le « Labour Left » au Royaume-Uni, les « Jusos » du SPD, ou la Confédération des syndicats suédois. Paradoxalement, on assiste dans le même temps à une technocratisation des enjeux européens, puisque ces réformes économiques sont surtout conçues par des experts socialistes alors favorables à une planification de l'économie par les pouvoirs publics, tels que le Britannique Stuart Holland, l'Italien Franco Archibugi, ou encore le Français Jacques Delors. Ces experts forment des réseaux transnationaux et rédigent plusieurs études au cours des années 1970 prônant la planification de l'économie, la redistribution des richesses et des revenus, et des nationalisations. Ces propositions ont cependant des difficultés à trouver un débouché politique, car les premières élections européennes de 1979 voient les partis socialistes se diviser. De son côté, la CES adopte une attitude combative à la fin des années 1970, mais sans grand succès.

Le sixième chapitre (« The Defeat ») aborde enfin l'échec des projets d'Europe sociale au début des années 1980. Des avancées ont bien été obtenues dans les années 1970, avec le renforcement des fonds de cohésion, l'adoption des premières directives sociales, et la multiplication des concertations avec les syndicats. Cependant, ces réalisations restent bien en deçà des revendications initiales de la gauche, comme le montre l'autrice à partir de deux exemples. Le premier est celui de la réduction du temps de travail : en dépit des premières manifestations européennes organisées par la CES, le Conseil n'adopte qu'une simple recommandation en juin 1984. Le second exemple est celui de la participation des travailleurs dans les multinationales. Le social-démocrate néerlandais Henk Vredeling, commissaire aux affaires sociales de 1977 à 1981, présente une proposition de directive en octobre 1980, mais celle-ci suscite une forte hostilité du patronat, redoublée par de vives oppositions au sein de la Commission et du Conseil. La directive est finalement enterrée par le Conseil européen en mai 1986, après avoir été vidée de ses ambitions au Parlement, où les eurodéputés socialistes n'ont pas su défendre le texte.

Dans l'épilogue de son ouvrage, Aurélie Dianara Andry dresse une synthèse des raisons pour lesquelles la gauche politique et syndicale européenne échoue à faire advenir les programmes d'Europe sociale qu'elle avait conçus. Certes, le cadre institutionnel et juridique des premières décennies de l'intégration européenne n'était pas favorable à des politiques sociales ambitieuses. Certes, le patronat européen a combattu vivement la mise en place de l'Europe sociale. Mais finalement, les raisons de cette défaite sont surtout à imputer à la gauche elle-même : celle-ci s'est montrée trop divisée, ses propositions programmatiques sont souvent restées vagues, les socialistes et les communistes n'ont pas su s'allier, et surtout, à l'exception de quelques tentatives sans lendemain de la CES, la gauche fut incapable d'initier de fortes mobilisations transnationales pour soutenir ses revendications. La route de l'Europe sociale ne fut pas prise par l'Union européenne, qui au contraire s'engouffra au cours de la décennie 1980 dans la route du néolibéralisme, sans depuis lors dévier de sa trajectoire.

Cette conclusion pousse le lecteur à se demander : la route de l'Europe sociale peut-elle de nouveau être ouverte, ou bien celle-ci s'est-elle refermée pour de bon ? Si l'autrice ne prend pas explicitement position, cet ouvrage rend plutôt enclin à opter pour la seconde réponse, d'abord parce que l'ensemble des conditions politiques et institutionnelles réunies dans les années 1970, et dont rend compte cet ouvrage, semble difficile à reproduire, et enfin parce que la gauche européenne paraît avoir largement abandonné l'Europe sociale comme un programme politique fort de réforme du capitalisme, pour à l'inverse la réduire à un simple rééquilibrage social du marché et de l'intégration européenne.

POLITIQUE EUROPÉENNE
N° 81 | 2023
Laura Chazel
p. 176-184

Giuliano Bobba et Nicolas Hubé, *Populism and the Politicization of the COVID-19 Crisis in Europe*, Londres, Palgrave Macmillan, 2021, 161 pages.

**Laura Chazel**
École des hautes études hispaniques et ibériques (EHEHI), Casa de Velázquez ; Univ. Grenoble Alpes, CNRS, Sciences Po Grenoble, Pacte, Grenoble, France

L'ouvrage collectif, dirigé par Giuliano Bobba et Nicolas Hubé, examine la manière dont les partis populistes européens ont réagi à la crise du Covid-19 pendant la première vague de la pandémie (mars –juin 2020). Ce travail se révèle précieux pour quiconque s'intéresse aux liens entre populisme et crise. L'ouvrage, auquel dix-neuf auteurs ont contribué, comprend dix chapitres, dont l'introduction et la conclusion et huit chapitres qui traitent chacun d'un pays. Ce travail collectif propose une analyse détaillée et rigoureuse des réponses apportées par les partis populistes (de gauche et/ou de droite) à la pandémie au Royaume-Uni[1], en Espagne[2], en Italie[3], en France[4], en Allemagne[5], en Hongrie[6], en Répu-

1 Ce chapitre s'intéresse principalement au discours de Nigel Farage, leader du *UK Independence Party* (UKIP, droite radicale) qu'il quitte en 2018 pour fonder le *Brexit Party* (parti du Brexit, droite radicale) à partir de 2019, dont il laisse la présidence en 2020.

2 Cas étudiés : *Unidas Podemos* (UP, gauche radicale) ; Vox (droite radicale).

3 Cas étudiés : *Movimento 5 Stelle* (Mouvement 5 étoiles, M5S, transversal) ; *Lega Nord* (Ligue du Nord, LN, droite radicale).

4 Cas étudiés : *La France insoumise* (LFI, gauche radicale) ; *Rassemblement national* (RN, droite radicale).

5 Cas étudié : *Alternative für Deutschland* (Alternative pour l'Allemagne, AfD, droite radicale).

6 Cas étudié : Fidesz (droite radicale).

blique tchèque[7] et en Pologne[8]. Chaque chapitre est divisé en quatre parties : (1) le contexte politique national ; (2) le niveau de circulation du virus pendant la pandémie et les mesures prises par le gouvernement au pouvoir pour freiner sa diffusion ; (3) le discours des partis populistes du pays étudié pendant la pandémie ; (4) la politisation du Covid-19 par les partis populistes.

Dans le chapitre introductif, Giuliano Bobba et Nicolas Hubé présentent le cadre théorique sur lequel repose l'ouvrage. Ils s'appuient sur une définition minimale du populisme en trois éléments (le peuple comme sujet du changement ; l'anti-élitisme ; et la désignation de « groupes marginaux » nuisibles) et caractérisent la crise comme « un moment où les contradictions systémiques [...] sont largement perçues comme saillantes dans les sphères politiques, culturelles ou idéologiques et où des interventions décisives et des transformations structurelles sont mises en œuvre » (p. 3). S'étayant sur ces concepts, l'ouvrage vise à répondre à deux objectifs. Premièrement, les auteurs analysent comment les partis populistes ont bénéficié de la crise sanitaire en examinant la manière (1) dont ils ont ajusté leurs discours à ce nouveau contexte et (2) dont ils ont politisé la pandémie. Pour examiner le processus de *politisation* de la crise, les auteurs mobilisent les travaux de William Felstiner, Richard Abel et Austin Sarat et s'appuient sur le triptyque « *naming* » (nommer le problème lors de son émergence), « *blaming* » (accuser les responsables du problème lors de la « phase de confrontation »), « *claiming* » (proposer des solutions dans la « phase de gestion » du problème)[9]. Deuxièmement, les auteurs analysent la manière dont la position institutionnelle occupée par les partis populistes (en opposition ou au pouvoir) a impacté leurs rhétoriques et leurs prises de positions.

Parmi les nombreux apports de l'ouvrage à la littérature existante, quatre méritent d'être soulignés. D'abord, l'ouvrage montre que les partis populistes au pouvoir ont tenté de dépolitiser la crise et ont profité de cette dernière pour montrer « leur capacité à prendre soin de la population même en cas de pandémie » (p. 134) – e.g. ce

7 Cas étudiés : *Komunistická strana Čech a Moravy* (Parti communiste de Bohême et Moravie, KSČM, gauche communiste) ; ANO 2011 (centre-droit) ; *Svoboda a přímá demokracie* (Liberté et démocratie directe, SPD, droite radicale).

8 Cas étudiés : *Konfederacja* (Confédération, droite radicale) ; *Prawo i Sprawiedliwość* (Droit et justice, PiS, droite radicale).

9 William L.F. Felstiner, Richard L. Abel, et Austin Sarat, « The Emergence and Transformation of Disputes: Naming, Blaming, Claiming... », *Law & Society Review*, 15 (3/4); p. 631-54, 1980.

fut le cas du *Movimento 5 Stelle* (Mouvement 5 étoiles, M5S, transversal) en Italie, de *Prawo i Sprawiedliwość* (Droit et justice, PiS, droite radicale) en Pologne, et d'*Unidas Podemos* (Unies, nous pouvons, UP, gauche radicale) en Espagne. Le cas du M5S est particulièrement intéressant compte tenu du fait que le parti n'a pas simplement tenté de dépolitiser la question sanitaire, mais a également inversé sa rhétorique : alors que ses leaders désignaient habituellement les « experts, les industries pharmaceutiques et les scientifiques » comme faisant partie de l'*establishment*, ils « sont rapidement devenus [pendant la pandémie] les seuls référents dignes de confiance pour mener les actions gouvernementales contre le virus » (p. 53).

Ensuite, concernant les partis populistes dans l'opposition, les auteurs montrent qu'ils ont, dans l'ensemble, continué à développer une rhétorique populiste – e.g. ce fut le cas en France, avec le *Rassemblement national* (RN, droite radicale) et *La France insoumise* (LFI, gauche radicale) qui ont « fortement politisé » (p. 67) la crise du Covid-19 en opposant le « peuple » aux « élites » incompétentes. Cependant, l'ouvrage met également en évidence que dans les premiers mois qui ont suivi l'apparition du Covid-19, les partis d'opposition ont soutenu les mesures de confinement, et n'ont pas élaboré de discours « anti-science » et « complotiste » – à quelques exceptions près, comme par exemple la *Lega Nord* (Ligue du nord, LN, droite radicale) en Italie (e.g. sur l'origine du virus ; sur le rôle de l'industrie pharmaceutique).

L'ouvrage révèle ensuite que les partis populistes ont, dans l'ensemble, utilisé la crise pour valider leurs propres idéologies. Les partis populistes de gauche (e.g. LFI, UP) ont développé une rhétorique mettant l'accent sur la nécessité de préserver des services publics robustes et sur la protection des plus démunis, tandis que les partis populistes de droite (e.g. Fidesz, RN, Vox) ont renforcé leur rhétorique eurosceptique et leur discours anti-migrants en opposant le peuple national, à sauver et protéger, aux immigrés (par exemple, le RN dressait une opposition entre « les seniors français et les immigrés », p. 66). Certains leaders populistes de droite se sont servis d'une rhétorique sur l'origine « étrangère » du virus (e.g. Javier Ortega Smith, secrétaire général de Vox, considérait les « "anticorps espagnols" comme moyen de combattre "ces maudits virus chinois" », p. 38), tandis que d'autres se sont étonnamment abstenus de le faire (e.g. Marine Le Pen, présidente du RN, déclarait le 29 février 2020 : « rien, absolument rien ne justifie que nos compatriotes d'origine chinoise soient victimes de défiance en raison du coronavirus. Ils n'y sont pour rien enfin ! Restons intelligents et fraternels ! », p. 65).

Enfin, l'ouvrage montre que, par sa dimension *sanitaire*, la crise du Covid-19 se distingue des autres crises qui ont touché l'Europe durant la dernière décennie (e.g. économique, « migratoire »). Selon les auteurs, puisque les facteurs à l'origine de la pandémie sont « exogènes, déterminés par des chocs extérieurs » (p. 7), les partis populistes n'ont pas pu politiser cette crise comme ils l'auraient fait habituellement – e.g. les élites responsables sont plus difficilement nommables puisque la pandémie fait partie des « événements indépendants de la volonté humaine » (p. 7). Dans le dernier chapitre, Giuliano Bobba et Nicolas Hubé – se fondant sur les huit cas traités dans l'ouvrage – concluent ainsi que, par sa particularité, la crise du Covid-19 n'a pas profité aux partis populistes comme cela est généralement le cas lorsqu'ils tentent de s'approprier les crises et « [d'alimenter] un "cycle de crise permanent" » (p. 142).

Plusieurs critiques, qui n'enlèvent rien à la pertinence scientifique de ce travail collectif, peuvent être adressées à l'ouvrage. Lorsqu'ils définissent le concept de « populisme », Giuliano Bobba et Nicolas Hubé s'appuient surtout sur le cadre théorique de l'approche idéationnelle[10]. Les représentants majeurs de cette approche qui sont cités dans le livre (e.g. Cas Mudde, Jan-Werner Müller, Cristóbal Rovira Kaltwasser) tendent à associer le populisme à un phénomène « illibéral » et « pathologique »[11], et développent une vision essentiellement péjorative du populisme (de gauche comme de droite) qu'ils considèrent principalement comme « l'autre » de la démocratie (libérale)[12]. La dimension moraliste et négative du populisme qui structure l'ouvrage collectif débouche, selon nous, sur quatre problèmes principaux.

Premièrement, cette vision dépréciative du populisme mène parfois les auteurs à affirmer des généralités aisément contestables. Les auteurs associent théoriquement, et à plusieurs reprises, le populisme (quelle que soit son idéologie hôte) au « complotisme » alors que les don-

10 Dans le champ académique, deux approches dominent aujourd'hui l'étude du populisme : l'école d'Essex et l'approche idéationnelle. L'école d'Essex a d'abord mis en avant le fait que le populisme (dans sa variante de gauche et progressiste) pouvait revitaliser la démocratie. Les directeurs de l'ouvrage collectif ne se réfèrent que brièvement aux travaux de l'école d'Essex et plus précisément à ceux d'Ernesto Laclau lorsqu'ils évoquent la relation entre populisme et crise (p. 4).

11 Federico Tarragoni, « La science du populisme au crible de la critique sociologique : archéologie d'un mépris savant du peuple », *Actuel Marx*, 54, p. 56-70, 2013.

12 Yannis Stavrakakis, « Populism, Anti-Populism and Democracy », *Political Insight*, 9(3), p. 33-35, 2018.

nées empiriques sur lesquelles ils s'appuient indiquent plutôt que cette association est largement erronée et que ce supposé lien théorique est questionné par la littérature existante, en particulier par les chercheurs de l'école d'Essex[13].

Dans le chapitre 6 sur l'Allemagne, Oliver W. Lembcke examine le discours de l'Alternative für Deutschland (Alternative pour l'Allemagne, AfD, droite radicale) et explique que « de nombreux populistes semblent remettre en question l'existence biologique du coronavirus, [alors que] l'AfD [ne l'a] pas remise en cause » (p. 82-83). Affirmer que les partis populistes ont nié l'existence du Covid-19 devrait être corroboré par des données empiriques, d'autant plus que l'ouvrage tend davantage à démontrer le contraire. Dans le contexte de la pandémie de Covid-19, certains auteurs de l'ouvrage associent également les partis populistes aux mouvements « anti-science » et « anti-masques ». Par exemple, dans le chapitre sur l'Espagne, Jaume Magre, Lluís Medir et Esther Pano affirment que le « discours anti-science et anti-expert [...] est typique [des] organisations [populistes] » (p. 43) tout en montrant dans le même temps que les partis « populistes » qu'ils étudient (Vox et UP) n'ont pas développé ce type de rhétorique. De leur côté, Osman Sahin et Bogdan Ianosev, traitant du cas britannique, considèrent que « contrairement à d'autres leaders populistes [...] Farage n'a pas rejeté les masques » (p. 28).

La grande majorité des partis considérés comme populistes n'ont pas critiqué le port du masque, alors que le cas suédois – bien « qu'exceptionnel » – montre que des acteurs politiques *mainstream* sociaux-démocrates ont estimé que le masque n'était pas forcément une mesure sanitaire efficace pour endiguer la propagation de la pandémie[14]. En effet, pendant la première vague, la politique sanitaire du premier ministre social-démocrate Stefan Löfven et d'Anders Tegnell, épidémiologiste en chef de la Suède en 2020, s'est structurée autour d'une stratégie « d'immunité collective ».

Seongcheol Kim s'est récemment intéressé à la « réduction conceptuelle du populisme » qui est opérée par l'approche idéationnelle qui consiste à le réduire à une idéologie qui opposerait un *peuple vertueux*

13 Voir, par exemple, Yannis Stavrakakis, « How did "populism" become a pejorative concept? And why is this important today? A genealogy of double hermeneutics », *POPULISMUS Working Papers*, 6. Disponible en ligne : <http://www.populismus.gr/wp-content/uploads/2017/04/stavrakakis-populismus-wp-6-upload.pdf>.

14 Liv Sunnercrantz, « Sweden », *In Populism and the Pandemic, A Collaborative Report*, édité par Giorgos Katsambekis et Yannis Stavrakakis, Populismus, 2020.

à une *élite corrompue* et « conspiratrice ». Il remet en cause les liens théoriques établis entre populisme et complotisme et rappelle que le fait « de parler au nom d'un "peuple" contre une "élite" [n'implique pas forcément] une vision manichéenne du monde qui soit conspirationniste »[15]. S'appuyer sur de tels travaux aurait pu permettre aux directeurs de l'ouvrage de ne pas s'étonner « [qu'] aucun des dirigeants [populistes] analysés [dans l'ouvrage] n'ait donné crédit aux théories du complot qui font généralement partie intégrante de la stratégie populiste, ou ne les ait soutenues » (p. 139).

Deuxièmement, le choix des cas examinés dans l'ouvrage est parfois discutable. Le lecteur peut se demander si les auteurs n'utilisent pas le terme de « populisme » pour désigner tout parti politique ou nouvelle entreprise politique nés en opposition au pouvoir établi et qui ne sont pas traditionnellement considérés comme des « partis de gouvernement ». Par exemple, dans le chapitre 8 sur la République tchèque, Ondřej Císař et Michal Kubát examinent le discours du parti de gauche radicale KSČM alors que la littérature le qualifie plutôt de « parti communiste réformateur »[16]. De la même manière, dans le chapitre 3 sur l'Espagne, Jaume Magre, Lluís Medir et Esther Pano considèrent *Unidas Podemos* comme un parti populiste qui aurait été « contraint de mettre l'accent sur [son] idéologie de gauche » (p. 31-32) pendant la pandémie du fait de sa position institutionnelle. On pourrait au contraire affirmer que, depuis 2019 au moins, la rhétorique populiste ne structure plus le discours d'UP[17]. En effet, il semblerait que depuis le départ d'Íñigo Errejón, principal « artisan » de la stratégie populiste de *Podemos*[18], en 2019, UP adopte un narratif de gauche beaucoup plus classique.

Troisièmement, l'ouvrage collectif souffre de l'absence d'une troisième question de recherche qui aurait pu être entièrement structurée autour de la question de l'influence de l'*idéologie hôte* sur le discours des partis étudiés. Les auteurs auraient ainsi pu rappeler plus explicitement que la littérature établit une distinction entre populismes inclusifs/de gauche et populismes exclusifs/de droite, et

15 Seongcheol Kim, « Taking Stock of the Field of Populism Research: Are Ideational Approaches 'Moralistic' and Post-Foundational Discursive Approaches 'Normative'? », *Politics*, 42(4), p. 492-504, 2022.

16 Luke March, *Radical left parties in Europe*, New York: Routledge, 259, 2011.

17 Voir, par exemple : Chazel, Laura, « Theory and practice of populism by the radical left: a comparative analysis of Podemos (2014-2020) and La France insoumise (2012-2020) », Thèse doctorale, Université Grenoble Alpes, 2021.

18 Kioupkiolis Alexandros, « Podemos: The Ambiguous Promises of Left-Wing Populism in Contemporary Spain », *Journal of Political Ideologies*, 21(2), p. 99-120, 2016.

prendre davantage en compte ces différences dans leurs analyses[19]. Dans le chapitre introductif, les auteurs abordent brièvement les différentes formes prises par le populisme (de gauche, de droite) et ambitionnent de « rendre compte des similitudes et des différences pour définir des modèles parmi les différents types de populisme (QR2) » (p. 11–12). Cependant, tout au long de l'ouvrage et dans le chapitre conclusif, les auteurs n'examinent que trop brièvement l'impact de l'idéologie des partis sur l'adaptation de leurs discours à ce nouveau contexte de crise sanitaire. Cela peut donner l'impression aux lecteurs que le concept de « populisme » est utilisé comme un terme « fourre-tout », ce qui mène les auteurs à dresser des parallèles parfois contestables entre populismes de gauche et populismes de droite.

Quatrièmement, les analyses proposées dans l'ouvrage auraient pu être davantage mises en perspective avec les discours des partis politiques considérés comme plus *mainstream*. À la lecture de certains chapitres, il semblerait que seuls les partis populistes tentent généralement de s'approprier les crises. Suivant les travaux de Yannis Stavrakakis, les auteurs auraient ainsi pu rappeler que les leaders populistes ne sont pas les « seuls à produire des récits de crise » et que ces récits « sont toujours confrontés aux discours dominants qui articulent leurs propres récits de crise »[20].

Les critiques adressées à l'ouvrage n'interrogent en rien sa pertinence scientifique, mais sont plutôt une invitation à ouvrir un débat sur la conceptualisation du populisme. Cet ouvrage collectif, que nous ne saurions trop recommander, est nécessaire et bienvenu dans la mesure où il est l'un des premiers à aborder la question de la réponse des partis populistes à la pandémie de Covid-19. Ce travail apporte une contribution importante à la littérature s'intéressant aux liens entre populisme et crise, un sujet qui, comme le soulignent les directeurs de l'ouvrage, a été sous-étudié en science politique. L'ouvrage contribue ainsi à rendre

19 Par exemple, les chercheurs de l'approche idéationnelle considèrent le populisme comme une idéologie (fine) incomplète à laquelle est toujours attachée une « idéologie hôte » (e.g. socialisme, libéralisme, conservatisme). Mudde et Rovira Kaltwasser (2017) distinguent ainsi (1) les populismes « inclusifs » (de gauche) qui délimitent la frontière entre le peuple et les élites à partir de critères socioéconomiques (e.g. les 1 % les plus riches contre le peuple) et qui permettent à des groupes exclus de participer au système démocratique ; (2) des « populismes exclusifs » (de droite) qui la délimitent à partir de critères socio-culturels (e.g. le peuple national contre les immigrés) et qui excluent certains groupes des ressources de l'État.

20 Yannis Stavrakakis, « Populism, Anti-Populism and Democracy », *Political Insight*, 9(3), p. 33-35, 2018.

plus intelligible la manière dont les partis populistes se nourrissent des crises, dont ils les politisent, et les usages qu'ils en font.

**Numéros parus :**

- **n° 1** 2000 La recherche en science politique et l'Union européenne
- **n° 2** 2000 Construction européenne et politiques sociales
- **n° 3** 2001 L'élargissement de l'Union européenne
- **n° 4** 2001 Mobilisations et clivages socio-politiques en Europe
- **n° 5** 2001 La Commission européenne
- **n° 6** 2002 Les partis politiques britanniques et l'intégration européenne
- **n° 7** 2002 Les groupes d'intérêt et l'Europe
- **n° 8** 2002 La défense en Europe
- **n° 9** 2003 Les parlementarismes européens
- **n° 10** 2003 L'Euro
- **n° 11** 2003 Administrer l'union européenne
- **n° 12** 2004 L'Europe au Microscope du Local
- **n° 13** 2004 Réformer l'Europe : la Convention européenne
- **n° 14** 2004 Enseigner l'Europe
- **n° 15** 2005 L'élargissement de l'UE – les nouveaux membres
- **n° 16** 2005 Européaniser les partis politiques
- **n° 17** 2005 À la recherche de la politique étrangère européenne
- **n° 18** 2006 La socio-histoire de l'intégration européenne
- **n° 19** 2006 Le patriotisme constitutionnel
- **n° 20** 2006 Genre et action publique en Europe
- **n° 21** 2007 L'Europe du chômage
- **n° 22** 2007 Sécurité extérieure de l'UE
- **n° 23** 2007 Sécurité intérieure de l'UE
- **n° 24** 2008 Dieu loin de Bruxelles
- **n° 25** 2008 Les approches sociologiques
- **n° 26** 2008 Amours et désamours entre européens
- **n° 27** 2009 Les syndicats à l'épreuve de l'Europe
- **n° 28** 2009 Les élections au Parlement européen
- **n° 29** 2009 L'Europe au miroir de la Turquie
- **n° 30** 2010 L'identité européenne, entre science politique et science-fiction
- **n° 31** 2010 Les effets de l'européanisation des politiques d'immigration
- **n° 32** 2010 Les groupes d'experts dans le gouvernement de l'UE
- **n° 33** 2011 L'intégration européenne par l'environnement. Le cas français
- **n° 34** 2011 Promouvoir l'Europe en actes
- **n° 35** 2011 Présider l'Union européenne
- **n° 36** 2012 L'espace européen du football
- **n° 37** 2012 D'une « mémoire européenne » à l'européanisation de la « mémoire »
- **n° 38** 2012 Les attitudes à l'égard de l'UE dans les nouveaux États membres
- **n° 39** 2013 L'Union européenne et le nouvel équilibre des puissances
- **n° 40** 2013 Les effets de l'européanisation dans le domaine social
- **n° 41** 2013 Varia
- **n° 42** 2013 Les politiques économiques européennes face à la Grande Récession
- **n° 43** 2014 Opposés dans la diversité
- **n° 44** 2014 Crise, crises et crisologie
- **n° 45** 2014 Les valeurs dans la gouvernance européenne
- **n° 46** 2014 L'exportation des normes au-delà de l'Union européenne

## Structures éditoriales du groupe L'Harmattan


**L'Harmattan Italie**
Via degli Artisti, 15
10124 Torino
harmattan.italia@gmail.com

**L'Harmattan Hongrie**
Kossuth l. u. 14-16.
1053 Budapest
harmattan@harmattan.hu

**L'Harmattan Sénégal**
10 VDN en face Mermoz
BP 45034 Dakar-Fann
senharmattan@gmail.com

**L'Harmattan Cameroun**
TSINGA/FECAFOOT
BP 11486 Yaoundé
inkoukam@gmail.com

**L'Harmattan Burkina Faso**
Achille Somé – tengnule@hotmail.fr

**L'Harmattan Guinée**
Almamya, rue KA 028 OKB Agency
BP 3470 Conakry
harmattanguinee@yahoo.fr

**L'Harmattan RDC**
185, avenue Nyangwe
Commune de Lingwala – Kinshasa
matangilamusadila@yahoo.fr

**L'Harmattan Congo**
219, avenue Nelson Mandela
BP 2874 Brazzaville
harmattan.congo@yahoo.fr

**L'Harmattan Mali**
ACI 2000 - Immeuble Mgr Jean Marie Cisse
Bureau 10
BP 145 Bamako-Mali
mali@harmattan.fr

**L'Harmattan Togo**
Djidjole – Lomé
Maison Amela
face EPP BATOME
ddamela@aol.com

**L'Harmattan Côte d'Ivoire**
Résidence Karl – Cité des Arts
Abidjan-Cocody
03 BP 1588 Abidjan
espace_harmattan.ci@hotmail.fr


## Nos librairies en France


**Librairie internationale**
16, rue des Écoles
75005 Paris
librairie.internationale@harmattan.fr
01 40 46 79 11
www.librairieharmattan.com

**Librairie des savoirs**
21, rue des Écoles
75005 Paris
librairie.sh@harmattan.fr
01 46 34 13 71
www.librairieharmattansh.com

**Librairie Le Lucernaire**
53, rue Notre-Dame-des-Champs
75006 Paris
librairie@lucernaire.fr
01 42 22 67 13

www.ingramcontent.com/pod-product-compliance
Lightning Source LLC
LaVergne TN
LVHW021208061026
849535LV00008B/2627
* 9 7 8 2 3 3 6 4 2 8 6 6 6 *